assessment and management

A field guide to the macrophytic vegetation of British watercourses

S.M. HASLAM
Botany School, University of Cambridge

P.A. WOLSELEY
Albro Castle, St Dogmaels, Cardigan

CAMBRIDGE UNIVERSITY PRESS

CAMBRIDGE
LONDON · NEW YORK · NEW ROCHELLE
MELBOURNE · SYDNEY

Published by the Press Syndicate of the University of Cambridge
The Pitt Building, Trumpington Street, Cambridge CB2 1RP
32 East 57th Street, New York, NY 10022, USA
296 Beaconsfield Parade, Middle Park, Melbourne 3206, Australia

© Cambridge University Press 1981

First published 1981

Printed in Great Britain at the University Press, Cambridge

British Library Cataloguing in Publication Data
Haslam, Sylvia Mary
River vegetation, its identification, assessment
and management.
1. Fresh-water flora — Great Britain — Identification
I. Title II. Wolseley, Patricia
581.9′29′41   QK306   79-41768
ISBN 0 521 23186 8 hard covers
ISBN 0 521 23187 6 paperback

Contents

Contents

Preface

This book identifies and classifies the plant communities found in British streams, dykes and canals, assesses damage and pollution, and gives the probable effects of other activities of man, from trampling to the construction of reservoirs. It therefore forms a manual for watercourse vegetation and its management. It is written for non-specialists — for anyone able to name plants, or able to use a simple book to do so. References are given to other sources of information.

Chapter 1 describes a simple method for surveying watercourse vegetation, and discusses identification. The procedure used for stream community classification is given in Chapter 2. Chapter 3 describes the principal stream communities found throughout Great Britain. Naturally, some local variants are not included, but these usually resemble, or can be deduced from the communities listed. Chapter 4 investigates the extent of damage in streams. This can be quantified in most lowland streams, and some hill ones, by comparing the actual vegetation found with that described in Chapter 3 as the 'correct' vegetation for the particular habitat. Damage can be physical (e.g. cutting, altering flow), or chemical (pollution). If there is little or no physical damage a pollution index can be used. Chapter 5 describes the vegetation of dykes and canals, and provides a rating scheme for these. The final chapter summarises the effects of different types of man's activities, providing a guide to the results of management and disturbance. It can be used both to predict, and to explain, changes in vegetation.

The vegetation described in Chapter 3 and parts of Chapter 5 is, for the purposes of this book, termed 'undamaged'. In Britain, only a few mountain streams have been unaffected by man: all lowland streams are managed, often intensively, and dykes and canals are man-made and depend on management for their existence. Further, much of Britain was originally covered by forest, which therefore shaded streams and confined their vegetation to intermittent openings in the tree canopy (and the central parts of larger rivers that could not be shaded). 'Undamaged' vegetation is, therefore, quite a different concept to 'natural' vegetation: natural vegetation is that which is unaffected by man, while undamaged vegetation is that with diversity, cover and quality as high as can be found in unpolluted British streams at the present time. The long-term effect of traditional management patterns is to maintain undamaged vegetation, though the short-term effects of these (e.g. dredging) and twentieth-century techniques (e.g. major re-channelling, herbicides) may be to cause damage.

Chemical assessments of pollution have been used in Britain since the early years of this century, with increasing sophistication. Invertebrate pollution indices, particularly the Trent Biotic Index, have been in use for several decades, and indices using other animal or plant groups (e.g. fish, algae) have occasionally been used (see J.M. Hellawell (1978). *Biological Surveillance of Rivers*. Water Research Centre). This book introduces plant (macrophyte) monitoring for Britain. There are several advantages in thus using the larger plants. First, botanists are more concerned with the effect of pollution on plants than on, say, fish. Also, if water is wanted for crop irrigation, then assessment by plants is the obvious choice. (However, even if potential irrigation water is clean on the plant index, pilot tests should be run on the crop(s) to be irrigated, as some crop species are exceptionally sensitive to pollution.) Another advantage is speed: macrophyte monitoring can be done in minutes, while invertebrate monitoring takes hours, and chemical assessments days. This is because the plants are easy to record, being both large and stationary (though they can be surveyed only in summer, as many die back in winter). When a river is recovering from pollution, plants can be valuable indicators of when water can be abstracted, because the plants are rooted in, and so affected by, the soil (the substrate) — and it takes longer to wash out polluted substrates than it does to substitute clean water. It is important that the substrate is clean if storm flow water is to be abstracted, because during storms silt from the bed is carried in the water, so if the silt is toxic the storm flow water will also be toxic. So even if the chemical and invertebrate indices report, quite correctly, that the water is clean, storm water should not be abstracted until the plant index reports the silt is clean also. Finally, and, for watercourse quality assessments, most importantly, it must be remembered that all the different animals and plants and (measured) chemicals reflect somewhat different aspects of that watercourse quality. Therefore it is only when each and all methods find a river clean that it can be assumed to be truly clean. If either the invertebrate or the macrophyte indices find the habitat polluted, it is so.

This book is for use in England, Scotland and Wales only. There is a general resemblance of vegetation types in different countries of Europe, but there is seldom an exact correspondence between, for example, sandstone streams in Belgium, Germany and Britain. The descriptions of communities given here therefore vary from being almost the same as those for watercourses outside Great Britain to being merely similar (and, as can be seen from Chapter 4 etc., a slight difference in the 'expected' (undamaged) vegetation renders the damage rating useless). The principles described here are international; the details, unless modified, are not. The authors hope to produce similar manuals for the other countries of the European Economic Community.

This research was funded by the Commission of the European Communities, under Contracts 079—74—1 ENV UK and 105—76—12 ENV UK, the Department of the Environment and the Natural Environment Research Council, whose support is gratefully acknowledged. It was based in Cambridge, and the help given by the Departments of Applied Biology (for contracts) and Botany (for facilities) is much appreciated.

We are much indebted to Dr F.H. Dawson, Mr S.G. Evans, Mrs M.P. Everitt, Dr A.J. Ferguson, Dr M. Furse, Mr R.M.A. Karim, Dr E.J.P. Marshall, Mr K.J. Murphy, Miss J. Pierce, Mr J. Terry, Mr D.F. Westlake and Mrs C. Wiggins for checking and improving the manuscript. Dr N.T.H. Holmes and Mr K.J. Murphy added extra species to the stream dial and canal list respectively. Mr S.G. Evans supplied some species lists for western clay streams, and Dr E.J.P. Marshall for New Forest streams. Miss M. Fowler amended the section on herbicides in Chapter 6. The rock type map was prepared by Mr P. Selden. The manuscript was typed by Mrs A. Hill. To all, we express our grateful thanks.

May 1979

S.M.H.
P.A.W.

I Surveying a watercourse

INTRODUCTION

There are as many ways of recording watercourse vegetation as there are botanists doing this work. The method described here is one of the simplest. It is advisable to follow the directions given for making field notes when using the book to classify and interpret vegetation, and it is *essential* to do so when diagnosing damage and pollution.

EQUIPMENT ETC. NEEDED

For identification

In the field

1. *British Water Plants*. S.M. Haslam, C.A. Sinker and P.A. Wolseley (1975). Field Studies Council.
2. Plant grapnel (Fig. 1.1) or heavily weighted hook on a long rope; polythene bag to keep this in when wet.
3. Polythene bags; in the car, buckets, sandwich boxes or equivalent containers to hold the polybags when they are filled with plants and water; in sunny or warm weather, a thick rug (or similar) to wrap over the containers for insulation.
4. Labels or scrap paper. Names written in ballpoint pen will remain legible in water for a few days.
5. Plant press. This can be made from two oven trays (wire-netting type) and two strong straps. Put newspaper 1–2 cm thick in the press, with some thin typing paper (copy paper) on the top. Place specimens between sheets of the typing paper, label them, and put them in the middle of the press, keeping new wet specimens separated from each other. If the car is being used regularly, plants will dry well there in summer, otherwise the press should be kept in a warm dry place until the plants are dry.

Fig. 1.1. Grapnel.

In the lab or at home

6. A.R. Clapham, T.G. Tutin and E.F. Warburg (1962). *Flora of the British Isles*, 2nd edn. Cambridge University Press.
7. 10X lens.
8. White dish in which living plants can be examined under water.
9. Thick white paper or thin card on which to mount pressed specimens. Plants may be attached by transparent tape or gum. Each should be labelled with the name of the plant, the site habitat and date of collection.
10. A reference collection of pressed specimens, prepared as above.

1 Surveying a watercourse

It is most important for beginners to be able to compare specimens with correctly identified ones. Even more experienced botanists can find reference specimens helpful with rarer species (e.g. grass-leaved *Potamogeton* spp.). Thirdly, if specimens of important plants are kept — whether these are common or rare — their identification can be checked if it is in doubt, and amended if necessary. Wrongly identified species can invalidate a whole research project.

With practice, the common plants can be identified while looking down from bridges. Beginners, and experts finding rare species, should collect specimens. The illustrations in *British Water Plants* show taxonomic features, and those in *River Plants* (see below) the characteristic appearance of the species in streams.

Equipment other than for identification

In the field

1. Notebook, and ballpoint pen or pencil; polythene bag large enough to hold the notebook when writing in the rain.
2. Gum boots or thigh boots.
3. Towel.
4. Spade or walking stick to help in going down slippery banks, walking in swift water on slippery stones, and testing the depth of loose mud on channel beds.
5. Measuring stick to test water depths. The rope on the grapnel, or the spade or walking stick, can be calibrated for this.
6. Quarter-inch (1 : 250 000) or larger scale Ordnance Survey maps of the area. The larger scale maps are required for alluvial plains, and are advisable if records are being taken very close together, and not from bridges, on streams. Non-O.S. maps should be used only if they show both small tributary streams and the National Grid.

In the lab or at home

7. S.M. Haslam (1978). *River Plants*. Cambridge University Press.
8. Cards, papers, files, etc., for the neat transcription of field records, and their basic interpretation.

TIME OF SURVEY

Between mid-June and mid-September all species are up, total cover is usually stable, and if the abundance of individual species is noted simply as 'much' and 'little', it also is usually stable. In southern England even late-growing species (such as *Sagittaria sagittifolia*) may be up by late May, and if it is known (from previous work) that only winter-green and early-growing plants are present, satisfactory records can be made in April. On the other hand, in cold years surveying in the north should wait until July. The first severe autumn storms usually disrupt stream communities. They typically occur in late September, but can vary from early September to, in mild and fine autumns, mid-October.

Surveys to diagnose damage and to identify communities must be done during this period of maximum growth. Seasonal changes, etc., are of course studied throughout the year.

Submerged plants cannot be seen in muddy or very swirling water so, after storms, surveying must wait until the water has cleared and the depth is nearly back to normal. If surveying is to identify the typical kind and amount of vegetation present, then it should be postponed for a few weeks after cutting, or after storms causing severe damage (even in the mountains, however, such storms are rare). Assessment of the damage from cutting, herbicides, storms, etc., should be done by surveys at frequent intervals, including (if possible) one before the event causing damage.

SITES FOR SURVEY

The species diversity data and damage ratings given in this book are for records from bridges or equivalent point sites (for the longer the reach of watercourse examined at any one place the greater, in general, is the number of species). For these purposes, therefore, *records must be taken from bridges or other point sites.* Wading in the watercourse to confirm identifications or look for small species must be confined to the area clearly visible from the bridges — say 10 m each side of the bridge.

It is important to look down into the water. Looking sideways means that submerged plants are often hidden by reflections or by emerged or floating plants. Except in shallow watercourses, therefore, recording should be done from bridges (unless the survey is important enough to justify diving or boating). Nearly all species can be seen and, with practice, recognised, from bridges. Wading along the sides of the watercourse may reveal extra species with unusually short (or hidden) shoots. Only a little experience is needed to find out which sites should be looked at in more detail; beginners should wade at each site where access and wading are possible.

If the water is rather turbid (not from storms) the plants that can be seen are normally the only ones present. In some streams, perhaps most in slow clay-mixed ones, small species such as *Potamogeton pusillus* may be found only by intensive search with a grapnel. As this is time-consuming, and the object of this book is to provide quick methods of identification, such species are not included in the lists of characteristic species in Chapter 3.

NUMBER OF SITES TO RECORD

The number of sites chosen depends on the purpose of the survey and the variability of the watercourse.

In order to familiarise themselves with watercourse vegetation, beginners should start not with a single site but with a river 12—20 miles (or 20—35 km) long, an equivalent length of canal, or a dyke (or dyke system) several miles (5—10 km) long.

For a general survey of a whole river on a single rock type and with typical topography, estimate the length, in miles, from the source to the mouth, and record this number of sites spread over both the main stream and the tributaries. If working in kilometres, the length of the river should be divided by 2 to get the number of sites. (Botanists surveying for Water Authorities must remember to record brooks as well as Main Rivers.) This will give a reliable diagnosis of stream type, and of upstream—downstream variation, but of course it may not include the effects of each effluent, reservoir, etc. With increased experience, and adequate information about effluents and management, fewer sites are needed for this basic classification.

To interpret a single uniform reach of a stream, three sites, all with similar vegetation, are needed at first (for a complete beginner, see above). With increasing practice this number can be decreased until reliable interpretations can be made from a single site, provided it contains much vegetation (two to three sites are necessary if there is little or no vegetation). Each site should be looked up separately in Chapters 2 to 4. In rapidly changing streams, where it is not possible to find three uniform sites (at suitable intervals), records from any one site are verified by those on either side. For instance, a stream with a bad effluent intake should show a sharp change from a clean to the uppermost polluted site, and then (if vegetation is quickly changing) a consistent improvement in the sites passing downstream — unless the vegetation is altered by some other damage factor such as dredging.

If the survey is for a specialised purpose, such as discovering the effect of one sewage works, the vegetation above the outfall should be assessed as described in the last paragraph, and that below should be recorded at intervals, the frequency of sites being anything from *c.* 20 m to *c.* 3 miles (*c.* 5 km) apart. If regular monitoring is required the number of sites chosen on each river (or canal etc.) can be small if the watercourse is clean, but in order to pick up slight changes in pollution as large a number of sites as possible should be recorded. A small change, such as the disappearance of blanket weed (long filaments of *Cladophora* etc.), is not significant from a single site, but is from seven consecutive sites spread over 5–20 miles (8–32 km).

RECORDING IN THE FIELD

For any site, records should begin with

1. Name of stream.
2. Name of nearest village or town.
3. Date of record.
4. Code number. If many sites are being recorded it is useful to give a code number to each. It is best to devise the code in relation to a river rather than to a county. It is convenient to give a basic number to each site. Then if more than one community is present, e.g. of fast and slow flow, these can be designated by (a), (b), etc. If sites are visited more than once, a further addition, perhaps in Roman numerals, can show the year of recording, or the number of times a site has been seen.

Next, the physical characteristics should be noted

5. Channel width in metres (approximate width where not affected by bridge).
6. Depth of most of the channel centre. With practise, this can be estimated by eye to the nearest 25 cm in waters up to 1 m deep, and to the nearest 50 cm in deeper places. An additional category, turbid water, may be needed when it is not practicable to measure depth.
7. The main types of substrate present, from the categories: boulder, stone, gravel, sand, silt and peat (mud is here included with silt). Note any particular patterns, e.g. silt bands towards the edges.
8. The main flow types present, from the categories:

Negligible	Water barely moving
Slow	Water obviously moving, water surface calm, and trailing plant parts still

Moderate	Water surface somewhat disturbed and swirling, trailing plant parts moving
Fast	Water surface disturbed, trailing plant parts moving vigorously
Rapid	Water surface broken by boulders or stones, much swirling and disturbance

9. The water clarity, from the categories:

Very turbid	Bed cannot be seen over 30 cm down
Turbid	Bed visible between 30 and 75 cm down
Clear	Bed visible at over 75 cm down
Shallow clear	Sites which appear clear, but are too shallow to determine whether they should be rated as somewhat turbid or as clear

Clarity varies far more than depth or flow type, and should be recorded on two separate occasions (neither shortly after storms) if it is suspected that it is an important factor.

10. The bank slope, both above and below the water surface, from the categories:

Gentle	less than 30°
Moderate	30°—60°
Steep	60°—90°
Vertical or undercut	

11. Possible sources of damage, for example: recent dredging, cutting, herbicides, boating, trampling, severe storm damage, etc.; much shade; piled sides; road works; obvious pollution. For full list see pp. 90—3.

12. Landscape. This is explained for beginners in Chapter 2.

Finally, the plants should be recorded

13. Plants above normal water level

The emerged water plants (those close to the water) should be recorded. These lists are rarely used for diagnosis, but can be helpful as additional evidence, e.g. during low flows.

The general habit of the bank vegetation should also be noted if the study is concentrating on the species of the water's edge — which can be suppressed by tall bank vegetation or intensive management. These habits include short (non-grass); grazed and ungrazed grasses; tall monocotyledons; tall rough (mixed dicotyledons); scrub; and trees. Other factors to record are any obvious management (grazing, cutting, herbicide, etc.) and how far the channel is sunk below ground level. Bank management is mainly outside the scope of this book, but is discussed briefly in Chapter 6.

14. Plants of the channel

All species present should be recorded. Fig. 1.2 defines which positions are to be considered as being in the channel and which on the bank. Some species occur in both habitats, and some are diagnostic of a vegetation type only when in the channel. For example, *Solanum dulcamara* is frequent on banks, infrequent in, though characteristic of, the vegetation of small chalk streams, and rare in most other stream types. *Epilobium hirsutum* is one of the commonest bank species, but as a channel species it is most frequent in smaller clay streams. *Myosotis scorpioides* is progressively restricted to the bank with increasing channel size and fluctuations of depth.

1 Surveying a watercourse

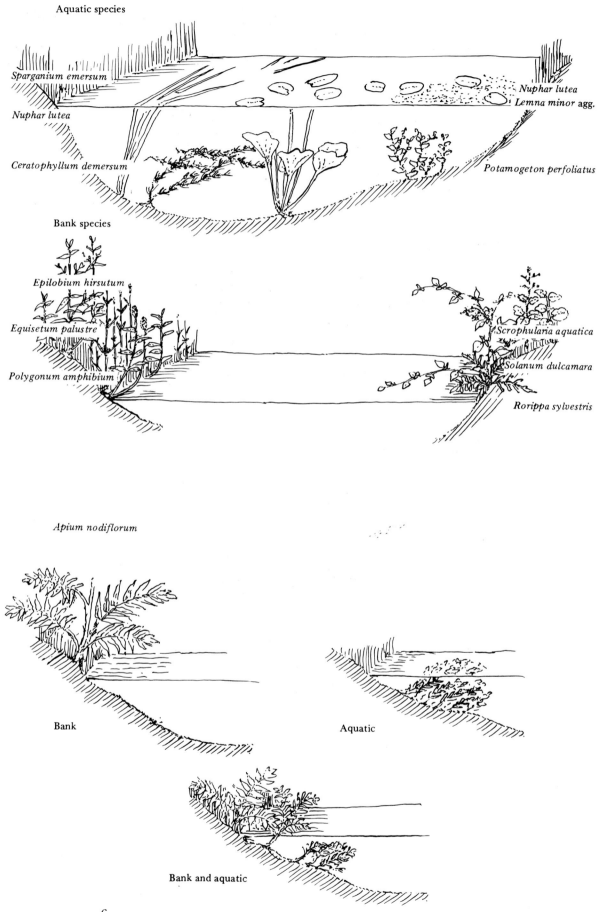

Aquatic species

Sparganium emersum

Nuphar lutea

Nuphar lutea
Lemna minor agg.

Ceratophyllum demersum

Potamogeton perfoliatus

Bank species

Epilobium hirsutum

Equisetum palustre

Polygonum amphibium

Scrophularia aquatica

Solanum dulcamara

Rorippa sylvestris

Apium nodiflorum

Bank

Aquatic

Bank and aquatic

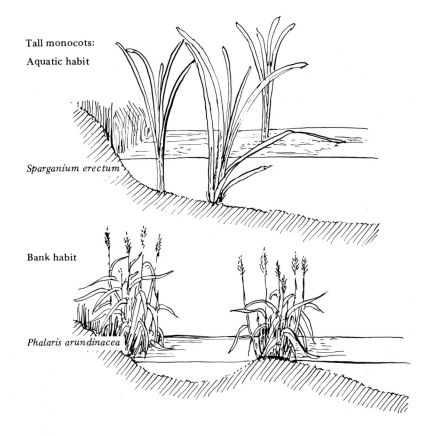

Tall monocots:
Aquatic habit

Sparganium erectum

Fig. 1.2

Bank habit

Phalaris arundinacea

Aquatic and bank habit

Epilobium hirsutum

The species listed below are normally bank species, not aquatics. If they, or species of similar habitat preferences, appear to be in the water, they should be carefully checked. Emergent plants must have their rooting base (not non-root shoots) below water level to qualify as growing in the channel.

Epilobium hirsutum
(except in small and clay-
based streams)

Juncus effusus

Lycopus europaeus

Polygonum amphibium
(when in the emergent
form)

Polygonum spp.

Rorippa islandica

Scrophularia aquatica

Solanum dulcamara
(except in small limestone-
based streams)

Symphytum officinale

These species vary in their habitat:

Carex acutiformis

Carex riparia

Phalaris arundinacea

Can often grow only above water level (except during storms)

Apium nodiflorum

Myosotis scropioides

*Rorippa nasturtium-
aquaticum* agg.

Veronica spp.

Can often grow only above water level in larger non-chalk rivers

Oenanthe crocata

As a water plant almost confined to nutrient-poor streams, but grows on the banks (and larger in size) in other stream types

Beginners should concentrate on the species in the 'Key to the Commoner Species of River Plants' in *British Water Plants*, and merely note the number of other species present and their habit (size, emerged or submerged, etc.). Plant shapes, and the way trailing plants move in currents are distinctive and should be memorised. The former are, in part, illustrated in Chapter 1 of *River Plants*. Plants which cannot be named from a bridge should be collected with a grapnel or by wading, and pressed specimens should be kept for future reference.

Two measures of plant abundance should be recorded.

(*a*) Each species should be given an abundance rating. If this is done on a two-point scale — little (L) or much (M) — then the records normally remain stable through the summer, and are likely to be the same when taken by different observers. A five-point scale is more flexible and allows comparisons between different months, different years, etc., but is likely to vary between observers and perhaps even within records made by the same observer. If sites are monitored regularly, more detailed comments are useful, such as 'abundant on silt bank to the left above the bridge', 'a 2 × 1 m clump *c*. 10 m beyond the left pier downstream'. (For an alternative method of listing species abundance see pp. 17—19.)

(*b*) The proportion of the watercourse covered by vegetation should also be noted — whether the plants are at the surface or on the bed. For streams, only plants in the shallower water are recorded: the percentage cover in water up to 1 m deep (or, in turbid or deep streams, that towards the sides of the channel). For dykes, drains and canals it is the total percentage cover that is wanted.

If *Lemna minor* agg. is important in a stream (as opposed to a dyke, drain or canal), the cover of this species should be recorded separately if a damage rating will be wanted. In dykes and drains (as opposed to streams or canals) the total plant cover, and the cover excluding tall edge plants, should be recorded for drainage ratings. In canals, the cover of submerged and floating plants within the bands at the sides (if such bands exist) is necessary for such ratings, as well as the total cover.

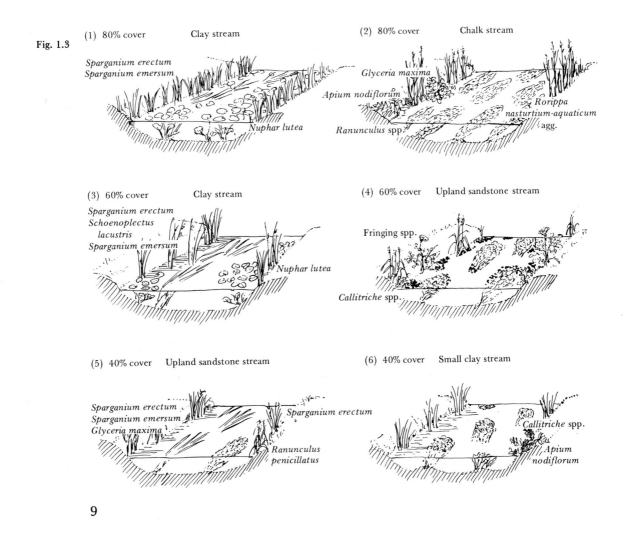

Fig. 1.3

(1) 80% cover   Clay stream

Sparganium erectum
Sparganium emersum

Nuphar lutea

(2) 80% cover   Chalk stream

Glyceria maxima

Apium nodiflorum

Rorippa nasturtium-aquaticum agg.

Ranunculus spp.

(3) 60% cover   Clay stream

Sparganium erectum
Schoenoplectus lacustris
Sparganium emersum

Nuphar lutea

(4) 60% cover   Upland sandstone stream

Fringing spp.

Callitriche spp.

(5) 40% cover   Upland sandstone stream

Sparganium erectum
Sparganium emersum
Glyceria maxima

Sparganium erectum

Ranunculus penicillatus

(6) 40% cover   Small clay stream

Callitriche spp.

Apium nodiflorum

9

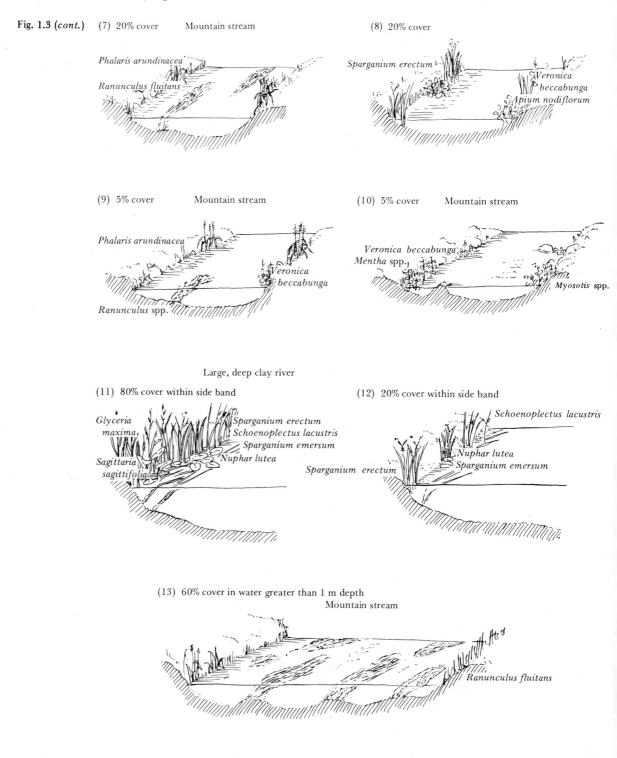

Fig. 1.3 (*cont.*)   (7) 20% cover   Mountain stream

Phalaris arundinacea

Ranunculus fluitans

(8) 20% cover

Sparganium erectum

Veronica beccabunga

Apium nodiflorum

(9) 5% cover   Mountain stream

Phalaris arundinacea

Veronica beccabunga

Ranunculus spp.

(10) 5% cover   Mountain stream

Veronica beccabunga
Mentha spp.

Myosotis spp.

Large, deep clay river

(11) 80% cover within side band

Glyceria maxima

Sparganium erectum
Schoenoplectus lacustris
Sparganium emersum
Nuphar lutea

Sagittaria sagittifolia

(12) 20% cover within side band

Schoenoplectus lacustris

Nuphar lutea
Sparganium emersum

Sparganium erectum

(13) 60% cover in water greater than 1 m depth
Mountain stream

Ranunculus fluitans

In larger streams bridges, or the weirs often associated with them, may alter the flow, depth, etc., of the channel. Where this occurs each main plant community should be recorded (e.g. in slow flow above the weir, very slow at the weir, fast below it, and slow again further from it). The community least influenced by the structure (weir, pier, etc.) is used for the assessment of pollution, and is the best for general identification of vegetation. The effect of the structure can be assessed from the difference between this community and those affected by the structure. In many streams, similar variations in vegetation occur naturally: for example in the alternating fast and slow reaches so common in mountain streams. Both communities should then be used in diagnosis.

Field diagnoses

At this stage the botanist can proceed to Chapter 2 (for streams) or 5 (for dykes, drains and canals), and classify and interpret the vegetation while in the field. As noted above, however, beginners should not place much weight on results from single sites. It is best to interpret in the field after visiting each batch of sites, but interpretation can also be done later in the lab or at home.

RECORDS TO BE COMPLETED IN THE LAB OR AT HOME

1. Plant identifications should be completed, and specimens wanted for reference should be pressed and, in due course, mounted. If living plants have to be kept for more than a day after collection they should be put in a fridge, still in their polythene bags.

2. Grid references should be looked up and noted for each site. For those working with large-scale O.S. maps, six-figure references can be given for each site, but four-figure ones are all that are actually needed for road bridges on streams outside towns.

3. If relevant, make further notes on disturbance and management, such as sites of effluents, date of last dredging, or of regulation of flow. This information should be sought first in the annual reports of the Water Authorities (and of their predecessors, the River Authorities and River Boards) and of the Scottish River Purification Boards. The Authorities themselves can be consulted if the information is not available elsewhere, but they cannot be expected to welcome trivial or frivolous enquiries.

4. Write out the complete record in a standard form and include on this, for streams, the landscape type and rock type (to be found from Chapter 2), and for dykes, the subsoil type (see Chapter 5). The records should be arranged so that each can be found easily. A card index is usually best, with each item in the same place on each card (e.g. grid reference in the lower left-hand corner).

5. Beginners, and all those concerned with whole rivers or with downstream changes, should construct river maps (see Chapter 11 of *River Plants*). A diagram — rough or exact depending on the purpose of the survey — should be made of the river, showing all streams marked on the quarter-inch (1 : 250 000) O.S. maps (as well as any others wanted). Then the plants found at each survey site should be marked. All other relevant detail should also be entered, e.g. variations in topography, rock type, stream size, and major effluent outfalls. Fig. 1.4 shows a rough (*a*) and an accurate (*b*) vegetation map of Upper R. Chelmer (Essex). The former is very quick to produce. Both maps show the vegetation type and the stream pattern, but the rough map does not locate individual sites.

6. Finish the interpretations of the data, using Chapters 2 to 5, if these were not completed in the field. If further understanding of the vegetation is wanted, study the relevant chapters of *River Plants*, the references listed therein, and later works as they become available.

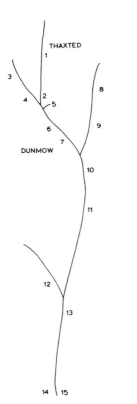

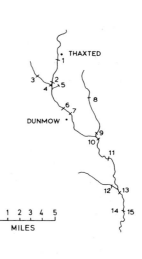

Fig. 1.4. Lowland clay stream, Upper R. Chelmer, Essex (1972): (*a*) diagrammatic, (*b*) accurate. Species present at the sites numbered: asterisks indicate luxuriance.

1. *Apium nodiflorum, Callitriche* sp., *Sparganium erectum.*
2. *Apium nodiflorum, Callitriche* sp., *Elodea canadensis, Epilobium hirsutum, Phalaris arundinacea, Rorippa nasturtium-aquaticum* agg., *Sparganium erectum.*
3. *Veronica anagallis-aquatica* agg.
4. Nil.
5. *Epilobium hirsutum, Rorippa nasturtium-aquaticum* agg.
6. *Agrostis stolonifera, Apium nodiflorum, Epilobium hirsutum, Phalaris arundinacea, Sparganium erectum, Zannichellia palustris*.*

7. *Elodea canadensis, Epilobium hirsutum, Phalaris arundinacea, Sparganium emersum, Sparganium erectum, Enteromorpha* sp., blanket weed*.
8. *Carex acutiformis, Iris pseudacorus, Potamogeton natans, Rorippa nasturtium-aquaticum* agg.
9. *Groenlandia densa,* blanket weed.
10. *Lemna minor* agg., *Potamogeton crispus*, Sparganium emersum*, Sparganium erectum*, Enteromorpha* sp., blanket weed.
11. *Callitriche* sp., *Potamogeton crispus, Potamogeton pectinatus*, Sparganium emersum, Zannichellia palustris*, Enteromorpha* sp., blanket weed.
12. *Apium nodiflorum*, Callitriche* sp., *Rorippa nasturtium-aquaticum* agg., *Veronica beccabunga.*

13. *Butomus umbellatus, Nuphar lutea, Potamogeton crispus, Sagittaria sagittifolia, Schoenoplectus lacustris, Sparganium emersum, Sparganium erectum.*
14. Millstream: *Agrostis stolonifera, Apium nodiflorum, Butomus umbellatus, Glyceria maxima, Phalaris arundinacea, Rorippa nasturtium-aquaticum* agg., *Sagittaria sagittifolia*, Sparganium erectum, Enteromorpha* sp.*, blanket weed

15. Main stream: *Agrostis stolonifera, Nuphar lutea, Phalaris arundinacea, Sparganium erectum*, Enteromorpha* sp., blanket weed*.

2 Identifying stream communities

INTRODUCTION

The stream dial (at the end of the book) identifies stream communities. (Canals, and the dykes and drains of alluvial plains, are described in Chapter 5.) The three white dials identify the habitat, and the vegetation appropriate to each combination is described in Chapter 3. The outer coloured dial identifies some characters of the plant community.

STREAM SIZE: INNER WHITE DIAL

'Water-supported species' are floating or submerged plants, including the larger algae and mosses but excluding small algae etc. (phyto plankton, benthic algae, periphyton and sewage fungus). However, small grasses (excluding *Glyceria fluitans* with long floating leaves) are not classed here, even when they are submerged.

'Spatey rivers' are those in the hills that are liable to severe storm flows able to cause substantial damage to vegetation.

    i.   Small streams without water-supported species
        c. 1–3 m wide. Either empty or with emergent aquatics.
    ii. Small streams with water-supported species
        c. 1–3 m wide. Water-supported species present, emergents present or absent.
   iii. Medium streams
        4–8 m wide. Any type of vegetation or empty.
    iv. Large non-spatey rivers
        Usually lowland, slow, deep and silting. Canalised rivers included here. 10+ m wide, usually with vegetation.
    v.   Large spatey rivers
        In highland regions. Often at least intermittently swift, shallow and stony, with or without vegetation.

Stream depths and further definitions are given under each habitat described in Chapter 3. A 3 m stream which is unduly deep for a small stream will bear vegetation typical of a medium stream (size iii), and should be looked up as a medium stream. Similarly a narrow and shallow medium stream is classed, vegetationally, as a small one (size ii), and the same applied to the division between medium and large streams.

If it is not clear whether a site should be classed as size iv (large non-spatey) or v (large spatey), there is normally only one of these listed under the relevant rock type.

13

LANDSCAPE TYPE: MIDDLE WHITE DIAL

The applicable landscape type is usually that of the upper part of the catchment, which can usually be identified by the pictures on the dial and from Table 2.1. (The applicable catchment is that for each stream as a whole. A river rising in the mountains, and mountainous in character throughout its length, can have both upland and lowland tributaries.) When in doubt, however, the topographical classes are as given in Table 2.1.

Table 2.1. *Landscape classification*

Topographical class	Hill height (usual)	Fall from hill top to stream channel in upper reaches (usual)	Slope of channel of upper streams (from one-inch map) usual	Liability to spate
Plain	—	—	—	Nil
Lowland	up to 800' (245 m)	up to 200' (60 m)	flatter than 1 : 100	Nil
Upland	800–1200' (245–365 m)	300–500' (90–150 m)	1 : 40 to 1 : 80	Some
Mountain	2000' (610 m) and more	600' (185 m) and more	steeper than 1 : 40	Much
Very mountainous	2000' (610 m) and more	1000' (305 m) and more or rainfall very high	steeper than 1 : 40	Great

The third column in Table 2.1, the fall from hill to channel, gives the most important figure, so a doubtful site should be classed using this information. Doubtful sites are found particularly where low (*c.* 1200':365 m) hills have very sharp falls from hill to channel (e.g. North Yorks Moors, mountain streams) and where flatter country is raised well above sea level (e.g. Dartmoor, lowland streams). Such sites can be looked up in the text under both possible landscape types, and the closest description taken.

The landscape types are divided according to land use as well as topography. There are two classes of plains, and two of lowland.

Plain A. Alluvial plains are flood plains with fertile alluvial soil, under grass or arable, around sea level (e.g. the Fenland, Romney Marsh, Somerset Levels).

Plain G,H. Blanket bogs or bog plains, occur in flat, high-rainfall hard-rock areas where acid (*Sphagnum*) peat has accumulated. Two stream types occur with blanket bog — small streams which are exclusively in the bog (H), and larger streams in country which is partly bog and partly hill (G).

Lowland B. Lowland farmland is the landscape type found in most of the south and east of England, and in the lower-lying and more fertile parts of the north and west. It is lowland in topography, fertile in soil type, and mostly arable or good-quality grassland in land use.

Lowland F. Lowland moorland and heath occur where the landscape is lowland even though, as on Dartmoor, it may be sited well above sea level, and the soil is acid peat (or podsol). The land is typically dominated by heathers (*Calluna* or *Erica*).

Additional notes

1. When a stream rises in one topographical type and flows on to another, the applicable type is usually the steeper of the two. A stream rising in the mountains and flowing to lowland or flood plain will remain mountainous on entering the flood plain. However, if this stretch is long in comparison with the stretch in the mountains the vegetation will gradually change to that of a quieter flow type (see Chapter 3, lowland streams rising in the hills: p. 50). The streams flowing in the alluvial plains, and classed as streams of plains, have all risen in the lowland (or upland) behind these: channels solely in alluvial plains are dykes and drains, which are considered separately (in Chapter 5).

Fig. 2.1

2. Soft limestone and soft sandstone (see rock type map, pp. 126—45) are always lowland, even when reaching *c.* 800' (245 m).
3. Clay, at 700' (215 m) and above, is upland.
4. Rainfall is higher in the west than in the east. This higher rainfall means that streams in hills in the west have a higher water force and, effectively, provide a habitat for the river plants equivalent to that of a steeper landscape type. For example in Devon, and the far north of Scotland, in particular, streams with upland (to lowland) topography can have vegetation appropriate to mountain streams. Unduly low rainfall similarly leads to a more lowland type of vegetation.

This point need be checked only if abnormal damage ratings occur in hill streams.

ROCK TYPE: OUTER WHITE DIAL

The map on pp. 126—45 can be used to identify rock type.

Additional notes

1. When a river flows over two or more rock types, the appropriate descriptions for streams on each of the types should be consulted, together with the general notes on mixed catchments (pp. 41—3, 79—81) and the description of the vegetation of the specific combination of rock types concerned.

2. Small outcrops cannot be shown on this small-scale map, nor can local variations in thickness or type of Boulder Clay. If the vegetation does not correspond to that given in the text, and the community does not appear damaged (this is described further in Chapter 4), a large-scale geological map should be consulted.

COLOUR BAND DIAL

The commoner or more diagnostic river plants are listed on this dial. Most are arranged in the colour bands. These species are roughly in the order of their nutrient status, from the nutrient-poor *Spaghnum* spp. to the nutrient-rich *Nuphar lutea*. Each species is classified according to its median habitat preference. Some species are confined to a very narrow habitat range (e.g. *Sphagnum* spp.). Other species have wider ranges, but do not occur with species far removed from the position at which they are shown (e.g. *Potamogeton alpinus, Rorippa amphibia*). *Sparganium erectum* is one of the most wide-ranging species, occurring with most of the other species listed, though with a preferred habitat centred on the position at which it is shown. Some species are listed at more than one position on the dial, which means they may either, like *Phalaris arundinacea*, also grow any-between the two habitats in which they are named, or, like *Nuphar lutea*, have quite a large habitat band in which they are absent or very rare. This list on the dial has been arbitrarily split into seven colour bands, each described by a name referring to nutrient status. (The *green* and the *blue* bands should in fact appear in parallel, but this is not possible on a flat dial.) Other species, which cannot be arranged in this way, appear on the *white* band. To assess species diversity, count all the aquatic species present, except benthic algae and sewage fungus. If land species are present, all count as one species. *Lemna minor* and/or *L. gibba* (not normally distinguishable in Britain) also count as one species, as do *Callitriche obtusangula, C. platycarpa* and *C. stagnalis*, and also all mosses. Only those species which are on colour bands, not those on the white band, are used to assess the colour band applicable to the site.

The colours are taken as being in the order *brown, orange, yellow, green, blue, purple, red*. (Put another way, the *white* band on the dial is a barrier that is not crossed.) Thus a community classed as *orange* to *blue* will have *orange, yellow* and *blue* species, and one classed as *blue* to *red, blue, purple* and *red* species.

Nuphar lutea   Should be classed in whichever colour band their main
Phalaris arundinacea   associates occur. (*Nuphar lutea* is very rare in the
                     orange band.)

Blanket weed is long trailing threads of *Cladophora* and equivalent algae. Lowland, and all abundant stands are *purple*. Sparse hill populations are *yellow*.

Mosses are considered as a single group, although there are many species with different habitat preferences. On hard rocks they are classed as *yellow*, on soft ones *blue*. (Hard rocks are rock types 7–11, soft ones types 1–6.)

Callitriche hamulata is *yellow*.

Callitriche spp. (not *hamulata*) are *blue*, and comprise *C. obtusangula, C. platycarpa* and *C. stagnalis*.

Glyceria fluitans with long floating leaves is *yellow*, but when in the commoner form, with shorter, emerged or semi-emerged leaves on plants fringing the banks, it is classed with other small grasses as *white*.

Catabrosa aquatica is the other small grass appearing on a colour band (*blue*). It has wider leaves than the others. If in doubt about identification, however, class as *white*.

Ranunculus spp. are notoriously difficult to identify, and habits and ecological preferences vary within species. *River Plants* should be consulted for illustrations and evidence. The comparative leaf length of *Ranunculus* spp. is from short to long, (a) *R. aquatilis, R. peltatus, R. trichophyllus*; (b) *R. penicillatus*; (c) *R. aquatilis, R. peltatus*; (d) *R. calcareus, R. penicillatus*; (e) *R. penicillatus*; (f) *R. fluitans*. Short-leaved *Ranunculus* which cannot otherwise be identified should be placed (from *yellow* to *purple*) with the other species from the site. Medium-leaved ones can be placed as *blue* with considerable certainty. *R. fluitans* is *green*, and fairly-long-leaved forms which are not *R. fluitans* (or its hybrids) are *purple R. penicillatus*. References to *Ranunculus* spp. are to Batrachian Ranunculuses. *R. flammula, R. sceleratus*, or land species are named separately.

Polygonum amphibium is counted only if it is in the floating form.

To assess the colour band of a site

1. Locate each species present at the site on the dial.
2. Ignore *white* species.
3. Decide which colour band, or two bands, best represent(s) these species. Abundant species should be considered more important than sparse ones.

Additional notes

4. A species list is rarely centred on the *green* band. However, if a *green* species is present in a list predominantly of another colour, the colour should be recorded as 'with *green*' (e.g. '*blue* with *green*').
5. If five or more 'coloured' species are present at a site, the colour band can normally be assigned with certainty.
6. If three or four 'coloured' species are present, and present in the same colour band, again this can be assigned with certainty.
7. If one or two 'coloured' species are present in one colour band, or three or four spread among several colour bands, only a dubious colour (*blue? orange—yellow?*) can be allocated to a single site, but when the same species/communities are recorded in many sites on the same river, the doubtful colour band can be confirmed.

Examples of deducing colour bands

(a = abundant, f = frequent, o = occasional, l = local)

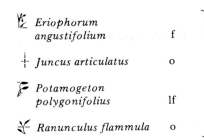

Eriophorum angustifolium	f	
Juncus articulatus	o	brown
Potamogeton polygonifolius	lf	
Ranunculus flammula	o	

⊽	*Callitriche hamulata*	o	⎫
✳	*Myriophyllum alterniflorum*	o	⎬ *orange–yellow*
⚘	*Phalaris arundinacea*	o	
✾	Mosses (on hard rock)	o	⎭

⊽	*Callitriche* sp.	o	⎫
♦	*Eleocharis acicularis*	lf	
‡	*Equisetum palustre*	o	
✳	*Myriophyllum alterniflorum*	o	
❦	*Nuphar lutea*	lf	
❦	*Nymphaea alba*	lf	⎬ *yellow* (with *green*)
⎘	*Potamogeton natans*	lf	
⎛	*Sparganium emersum*	o	
⚉	*Sparganium erectum*	lf	
✾	Mosses (hard rock)	o	⎭

⚘	*Phalaris arundinacea*	o	⎫
✴	*Ranunculus* sp. (short-leaved)	lf	⎬ *yellow* (?)
✾	Mosses (hard rock)	o	⎭

⚇	*Agrostis stolonifera*	o	⎫
❧	*Mimulus guttatus*	o	
⚇	*Myosotis scorpioides*	o	
✺	*Ranunculus trichophyllus*	f–a	⎬ *blue*
❀	*Rorippa nasturtium-aquaticum* agg.	lf	
✾	Mosses (hard or soft rock)	o	⎭

⊽	*Callitriche* sp.	o	⎫
✳	*Elodea canadensis*	o	
⚘	*Phalaris arundinacea*	o	
❧	*Rorippa amphibia*	o	⎬ *purple*
⎛	*Sparganium emersum*	o–f	
✾	Mosses (soft rock)	o	⎭

Colour band dial

Acorus calamus	o	
Butomus umbellatus	o	
Carex acutiformis	lf	
Lemna minor agg.	lf	
Nuphar lutea	a	
Potamogeton perfoliatus	lf	*purple–red*
Schoenoplectus lacustris	o	
Sparganium emersum	f–o	
Sparganium erectum	lf–f	
Typha latifolia	o	
Enteromorpha sp.	o	
Blanket weed	f	

Glyceria maxima	lf	
Nuphar lutea	f–a	
Sagittaria sagittifolia	la	*red*
Schoenoplectus lacustris	o–f	
Sparganium emersum	f	

Agrostis stolonifera	
Veronica beccabunga	see over

Phalaris arundinacea	
Mosses (hard rock)	see over

Callitriche sp.	
Mosses (hard rock)	see over

Phalaris arundinacea	
Ranunculus sp. (short-leaved)	see over

Agrostis stolonifera	
Callitriche sp.	see over

Mimulus guttatus	
Mosses (hard rock)	see over

19

If only one of the last six sites were recorded the colour band assigned would be *yellow?* or *blue?*, but with much doubt about its validity. If all six sites were recorded together, however, the sites differing somewhat in size but otherwise being uniform, the habitat could be assigned the colour *yellow—blue*.

DIRECTIONS FOR USING THE STREAM DIAL

1. Find the correct section on the inner white dial.
2. Align this with the correct section of the middle white dial.
3. Align these two with the correct section of the outer white dial.
4. Using the information on these three dials, look up the page number of the appropriate vegetation type in Table 2.2. Turn to the description on that page. Look up the expected colour band(s).
5. Align the three white dials with this colour band(s).
6. Check the list of species present at the side against those listed on that colour band of the dial.
7. Check the list of species present and the cover found against those listed as expected for that stream type.
8. (Where the community differs from that listed as expected in Chapter 3, consult Chapter 4.)

If there is some doubt over which landscape type (A—H) or which stream size (i—v) is applicable, descriptions of both possibilities should be read and the one closest — in habitat as well as vegetation — to the site in question should be chosen. If there is some doubt about which rock type is applicable (and a large-scale geological map is not immediately available) again look up the different possibilities to find that with the habitat description that best fits the site.

In descriptions of colour bands the terms used can include e.g. '*blue—purple*' or '*blue* to *purple*'. The former means the band is intermediate between *blue* and *purple* and the latter that the vegetation can range from completely *blue* to completely *purple*. Similarly '*blue* (to *purple*)' means that *blue* is more common but that *purple* is still within the normal range. Naturally the colour bands given are the characteristic ones and, as stated elsewhere, not the only ones found.

(Here, and elsewhere in this book, the term nutrient preference is a simplification. Species are affected by many habitat factors, and one 'preferring' a certain colour-band, may in fact be controlled by other habitat factors which, in Britain, happen to be correlated with nutrient status.)

Table 2.2. *Dial Directory*

1	B	i, ii, iii, iv	**Chalk**	*blue* to *purple*	p. 23
2	B	i, ii, iii	**Oolite**	*blue* to *purple*	p. 26
3	B, (D), C	i, ii, iii, iv	**Clay** (including upland)	*blue, purple, red*	p. 29
4	B	i, ii, iii (−iv)	**Soft sandstone**	*blue—purple* with *green*	p. 36
5	B	ii, iii	**New Forest sands**	*orange, purple,* with *green*	p. 38
6	A	(ii), iii, (iv)	**Alluvial**	*turquoise/blue* to *red* (see Chapter 5 for *turquoise*)	p. 40
7	B, D, E, F, G, H		**Resistant rock**		p. 43ff
7	H	i, ii	blanket bog, bog plains	*brown (brown—orange)*	p. 43
7	G	iii, v	rivers with much blanket bog	*brown, brown—orange*	p. 45
7	F	(i), ii, iii, (iv/v)	lowland moorland	*orange* to *blue*	p. 46
7	B	i, ii, (iii), iv/v	lowland farmland	*yellow* to *purple,* with *green*	p. 48
7	C	i, ii, iii, v	upland	*yellow* to *blue—purple*	p. 51
7	D	i, ii, iii, v	mountain	*blue* to *blue—purple*	p. 56
7	E	i + ii, iii, v	very mountainous	*orange—yellow* to *yellow* (with *green*)	p. 59
8	C, D, E (B, G, F)		**Hard sandstone**		p. 61ff
8	C, (B)	i, ii, iii, v	upland (and lowland)	*blue* to *purple—red*	p. 62
8	D	i, ii, iii, v	mountain	*yellow* to *blue—purple*	p. 65
8	E	i + ii, iii	very mountainous	*yellow* to *blue*	p. 67
8	B, F, C	i, ii, iii, (v)	Caithness	*yellow* to *purple*	p. 68
9	(B), C, D, (E)		**Hard limestone**		p. 69ff
9	C, (B)	i, ii, iii, (v)	upland (and lowland)	*(orange) yellow* to *blue—purple* (with *green*)	p. 70
9	D, (E)	(i), ii, iii, v	mountain (and very mountainous)	*yellow* to *blue,* with *green*	p. 74
10	C, (B)	i, ii, iii, (v)	**Calcareous and fell sandstone**	*blue* to *purple* with *green*	p. 76
11	C, D	i, ii, iii, v	**Coal Measures**	*yellow* to *purple* with *green*	p. 78

3 Descriptions of stream communities

Use a magnifying glass to identify the species in the illustrations in this chapter.

INTRODUCTION

This chapter describes the characteristic plant communities of the main stream habitats of Britain. Naturally, intermediates of communities and habitats occur, and where these are suspected both alternative habitats should be consulted. Not all sites will have the exact vegetation listed here. If undamaged, however, the vegetation should have the same diversity, colour band and percentage cover (see Chapter 1) as given here. These three factors are all used in the assessment of the damage rating in Chapter 4.

The layout is somewhat similar to a flora. Each section, divided according to rock type, is headed by a description applicable to all streams on that rock type (equivalent to a family description). Where more than one landscape type occurs on a given rock type the next subdivision is by landscape (equivalent to a genus). Finally, each size of stream found in that landscape and rock type is described and defined, and its typical vegetation is listed. When the species list is divided into abundant and associate species, all species occurring in the lists of abundants can be deemed to occur also in the list of associates.

Standard information given, for each type, includes:

Rock type ⎫
Landscape type ⎬ defined in Chapter 2 (see map on pp. 126–45 for distribution
Stream size ⎬ of rock types).
Colour band ⎭

Species diversity number: minimum number of species expected in an undamaged site (as defined in Preface, p. vii; Chapter 2, p. 16).

Percentage cover: minimum percentage cover expected in an undamaged site in water up to 1 m deep (or at sides, if all of river deeper or too turbid to see depth).

Cross-section illustration: this shows the typical appearance of the stream and its banks. The banks at the site being examined should be carefully compared with those in the illustration, since unusually steep slopes around water level decrease emergent vegetation, and unusually gentle ones increase it. These factors should be considered before damage is assessed (Chapter 4).

The 'fringing herbs' are the short semi-emerged dicotyledons frequently fringing small streams, i.e.

Apium nodiflorum   *Mentha aquatica*

Berula erecta   *Mimulus guttatus*

‍ ‍ ‍ *Myosotis scorpioides* ‍ ‍ ‍ ‍ *Veronica anagallis-*
‍ ‍ ‍ ‍ ‍ ‍ ‍ ‍ ‍ ‍ ‍ ‍ ‍ ‍ ‍ ‍ ‍ ‍ *aquatica* agg.
‍ ‍ ‍ *Rorippa nasturtium-*
‍ ‍ ‍ ‍ *aquaticum* agg. ‍ ‍ ‍ ‍ *Veronica beccabunga*

Unless there is a drastic change in climate the 'probable species' and colour bands should remain stable. Cover should remain stable in the lowlands, and is noted as variable in hill streams anyway.

The diversity data in this chapter come from records made in the mid- and late-1970s, when diversities were high following and during a period of low rainfall. It is likely that, over the lifetime of this book, rainfall and, consequently, diversity will alter. Changes in diversity from this cause will be observed by experienced and long-term workers and the book can be appropriately corrected. Changes will be less easily detected by beginners and short-term workers — and, as will be seen in Chapter 4, assessments of pollution depend on accurate expectations of diversity. If such a change is suspected, and no long-term workers are available for consultation, we recommend study of reaches of rivers whose history for the past eight years is known, and which are not suffering from pollution or physical damage. These survey records can be used as the standards.

Water depths may also alter if rainfall changes.

CHALK STREAMS

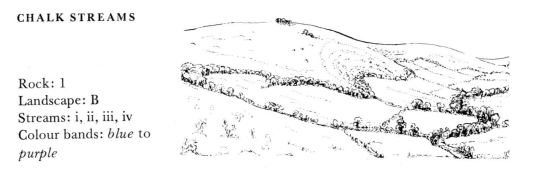

Rock: 1
Landscape: B
Streams: i, ii, iii, iv
Colour bands: *blue* to
purple

Chalk is a very pure soft limestone, normally forming rolling lowland hills up to *c.* 800′ (245 m). Much of the brook water comes from springs, giving fairly swift stable flows of sparkling clear water over gravelly beds with — for lowlands — little silt. The substrate is mesotrophic, with a high hardness ratio and particularly low magnesium, phosphate-phosphorus, chloride and sulphate-sulphur.

Chalk streams occur east from Dorset to Kent, and north from Hampshire to Yorkshire. The rock type must be checked with care, as many streams which are chalk streams in terms of fisheries or geography, e.g. R. Dorset Avon, have outcrops of other rocks in the catchment, which influence the vegetation.

i. Small streams without water-supported species

blue
3+ species
20+% cover

23

Local. 1—3 m wide, dry for most of the summer or for most or all the year.

Probable species: Mentha aquatica ① Veronica beccabunga ③

      Myosotis scorpioides ② Small grasses

      Phalaris arundinacea ④ Land spp.

      *Veronica anagallis-
      aquatica* agg.

If wetter: *Apium nodiflorum*

Wetter again: *Rorippa nasturtium-
      aquaticum* agg.

(a)

Variants:
(a) Dried grazed
channels (Fig. *a*)
(b) Sunk ditches (Fig. *b*)

(b)

ii. Small streams with water-supported species

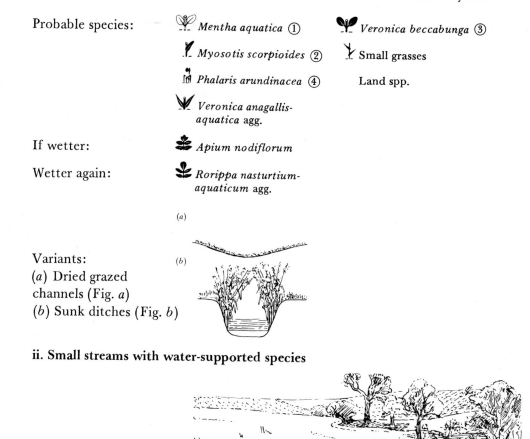

blue
4+ species
60+% cover

Common, 1—4 m wide, *c.* 20—40 cm deep, perennial flow or dried in late summer. Gravelly. *Apium nodiflorum* and *Berula erecta* frequently submerged. Fringing herbs in carpets, short bands or short small clumps (the carpets typically submerged, and even the clumps often partly submerged).

Often abundant: *Apium nodiflorum* *Ranunculus* spp.
              (short-leaved) ①
      Berula erecta ②
               *Rorippa nasturtium-
      Callitriche spp.   aquaticum* agg. ③

Often associated: *Catabrosa aquatica* ④ *Phalaris arundinacea*

      (*Hippuris vulgaris*)   *Sparganium erectum*

      Lemna minor agg.   *Veronica anagallis-
               aquatica* agg.
      Mentha aquatica
               Veronica beccabunga
      Mimulus guttatus
               Small grasses
      Myosotis scorpioides
               Mosses

Variant:

(*a*) *Potamogeton lucens* in flatter areas with some alluvium

iii. Medium streams

blue, blue—purple
6+ species
80+% cover

Common, 4—8 m wide, *c.* 30—75 cm deep. Flow usually moderate. Gravelly, little silt. *Apium nodiflorum* and *Berula erecta* frequently submerged. Fringing herbs in carpets, short bands, or short small clumps (the carpets typically submerged, even the clumps often partly submerged).

Often abundant: *Ranunculus* spp. (medium-leaved, particularly *R. calcareus*) ①

Smaller or shallower: *Berula erecta*

Callitriche spp. ②

Rorippa nasturtium-aquaticum agg. ③

Often associated: *Apium nodiflorum*

Carex acutiformis agg. ⑥

Catabrosa aquatica

Mentha aquatica

Myosotis scorpioides

Phalaris arundinacea ④

Sparganium erectum ⑤

Veronica anagallis-aquatica agg.

Veronica beccabunga

Small grasses

Mosses

And in deeper slower siltier reaches, one to three of:

Elodea canadensis

Glyceria maxima

Groenlandia densa

Oenanthe fluviatilis

Schoenoplectus lacustris

(*Sparganium emersum*)

Blanket weed

Note. In swifter parts the species in the last list indicate eutrophication.

iv. Large streams

blue—purple, purple
9+ species
80+% cover

Infrequent. 10+ m wide, 75 cm or more deep. Flow usually moderate to slow. Some silting. Fringing herbs usually in wisps or small short clumps.

Often abundant:
Ranunculus spp. (long—medium-leaved, particularly *R. calcareus*) ①

Very shallow: *Berula erecta*

Very silty: *Schoenoplectus lacustris* ②

Often associated:

Apium nodiflorum

Callitriche spp.

Carex acutiformis agg. ③

Elodea canadensis ④

Glyceria maxima ⑤

Mentha aquatica

Oenanthe fluviatilis ⑥

(Potamogeton crispus)

Rorippa nasturtium-aquaticum agg.

(Rumex hydrolapathum)

Sparganium emersum ⑦

Sparganium erectum ⑧

Veronica anagallis-aquatica agg.

Zannichellia palustris ⑨

Small grasses

Blanket weed

OOLITE STREAMS

Rock: 2
Landscape: B
Streams: i, ii, iii
Colour bands: *blue, blue—purple, (purple)*

Oolite is a soft limestone, less pure than chalk, normally forming rolling hills up to *c.* 400—800′ (120—245 m). Oolite usually outcrops on ridges, with clay (solid or Boulder) in the valleys between and sometimes also on the ridge. Oolite streams therefore soon flow on to clay, and there are no large and few medium streams solely on oolite. Brook flows vary from chalk-like (swift stable flows of sparkling clear water over gravelly beds) to slower, less clear, and more silty streams. As on chalk, the substrate is mesotrophic with a high hardness ratio and low nutrient levels.

Oolite streams occur from Somerset to Lincolnshire, being mostly in the Cotswolds, Northamptonshire Wolds, and the ridges projecting north from these into Lincolnshire (the Coralline oolite of the North Yorks Moors is here classed as hard limestone and, being with sandstone, as limestone/sandstone for the area as a whole).

i. Small streams without water-supported species

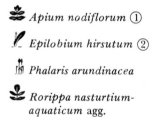

1 *blue*
   3+ species
   40+% cover

Local. 1–3 m wide, dry for most of the summer.

Probable species: Apium nodiflorum ①   Sparganium erectum ③

                       Epilobium hirsutum ②   Veronica beccabunga

                       Phalaris arundinacea   Grasses

                       Rorippa nasturtium-   Land spp.
                       aquaticum agg.

Variant:
(*a*) Dried grazed channels (see Fig. (*a*) on p. 00)

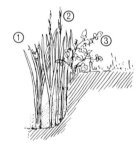

2 *purple* with tall
   emergents only
   1+ species
   20+% cover

Local. 1–3 m wide, dry for most of the year. Often sunk with steep banks

Probable species: Sparganium erectum ①

Often associated: Epilobium hirsutum   Solanum dulcamara ③

                       Glyceria maxima ②   Land grasses

                       Phalaris arundinacea   Land spp.

ii. Small streams with water-supported species

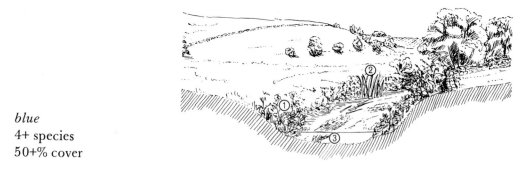

blue
4+ species
50+% cover

Frequent. 2–4 m wide, usually 20–30 cm deep. Moderate flow. Silty and gravelly.

Often abundant:	*Apium nodiflorum* ①		*Sparganium erectum* ②
	Ranunculus spp. (short- and medium-leaved) ③		Blanket weed
Often associated:	*Berula erecta*		*Sparganium emersum*
	Callitriche spp.		*Enteromorpha* sp.
	Phalaris arundinacea		and one to three others from the *blue* or *purple* bands

Variants:
(*a*) Brooks indistinguishable from chalk (e.g. upper R. Thames)
(*b*) Brooks like chalk–clay ones (see below)
Note. With variant (*b*) eutrophication as a result of man's activities can be diagnosed only if there is, in the vicinity, an otherwise similar stream (or a reach of the same stream) with a species assemblage from a *bluer* band. Then, the difference between the two indicates the degree of eutrophication.

iii. Medium streams

purple
5+ species
50+% cover

Infrequent. (3–)4–8 m wide, *c.* 30–75 cm deep. Flow usually moderate. Some silt.

Often abundant:	*Ranunculus* spp. (short- or medium-leaved) ①		*Sparganium emersum* ②
			Blanket weed

Clay streams

Often associated: 🌿 *Apium nodiflorum* ②   〜 *Enteromorpha* sp.

⫟ *Schoenoplectus lacustris* ③   Often other, usually

𝕎 *Sparganium erectum* ④   *purple*, spp.

OTHER SOFT LIMESTONES

Small outcrops of other soft limestones (e.g. in Corallian) are not large enough to bear streams solely on limestone. See under soft limestone—clay streams below.

CLAY STREAMS

Rock: 3
Landscapes: B, C
Streams: i, ii, iii, iv
Colour bands: *blue* to *red*

Clay is the commonest rock of lowland Britain. It strongly influences river vegetation, and clay communities also occur where thick Boulder Clay overlies other rock types and when small outcrops of other rocks occur in a mainly-clay catchment. Brook water comes from run-off and so is unstable, depending on rainfall. The non-porous land surface increases the instability of flow, and leads to there being many brooks in a given area. Much silt is washed off the land onto the stream bed and this, coupled with the unstable flow, leads to substrates which are silty and, especially in river centres, unstable. The water is usually somewhat turbid. The substrate is eutrophic, both because much silt is present and because the silt is nutrient-rich, with the highest levels of calcium, magnesium, sodium, phosphate-phosphorus, chloride and sulphate-sulphur found in any stream type and a medium hardness ratio.

CLAY STREAMS: B LOWLAND

Rock: 3
Landscape: B
Streams: i, ii, iii, iv
Colour bands: *blue* to *red*

The landscape varies from nearly flat to low hills up to *c.* 600′ (185 m). The low topography leads to slow flows and much silting.

Lowland clay is the commonest habitat in southern and eastern England, and (as thick Boulder Clay) is locally important elsewhere (e.g. Anglesey). For northwest lowland (upland) streams, see the section on upland and mountain streams below (p. 34).

3 Descriptions of stream communities

i. Small streams without water-supported species

1 With tall emergents only
blue—purple
1+ species
30+% cover

Frequent. 1—3 m wide. Dry for part of the year (or shallow with very slow flow). Often sunk *c.* 2 m below ground level, with steep banks, the tall plants of which may partly shade the channels. Substrate usually silty. Often shaded with little or no vegetation.

Often abundant: drier	*Epilobium hirsutum* ①	Land grasses
	Urtica dioica	Land spp.
wetter	*Sparganium erectum* ②	
Often associated:	*Glyceria maxima* ③	*Phalaris arundinacea*

Variant:
(*a*) Grade into ditches

2 With short emergents
blue with *Epilobium
hirsutum*
3+ species
10+% cover

Frequent. 1—3 m wide, shallow (e.g. 5—20 cm), sometimes dry for part of the summer. Slow to moderate flow. Silty to gravelly. With more scour than (1), sometimes less deeply sunk, and often somewhat further downstream. Fringing herbs often tall, sometimes in large clumps. Many shaded, with little or no vegetation.

Often abundant:	*Apium nodiflorum*	*Sparganium erectum* ①
Often associated:	*Epilobium hirsutum*	*Rorippa nasturtium-aquaticum* agg. ②
	Myosotis scorpioides	*Veronica beccabunga*
	Phalaris arundinacea ③	Small grasses

Clay streams

Variants:
(*a*) Dry with much grass, sometimes grazed
(*b*) Swift flow with coarse substrate and little vegetation

ii. Small streams with water-supported species

1 Smaller
   blue
   2+ species
   20+% cover

Common. 2—3 m wide, shallow (*c.* 10—30 cm), sometimes dry for part of the summer. Moderate to slow flow. Silty to gravelly. Often sunk 1—3 m below ground level and partly shaded by tall plants on these high banks. Fringing herbs often tall, sometimes in large clumps, always emerged. Many shaded, with little or no vegetation.

Probable species: *Apium nodiflorum*   *Sparganium erectum* ①

Callitriche spp. ②   *Veronica beccabunga* ③

Rorippa nasturtium-aquaticum agg.

If vegetation is less than expected, this is probably due to causes other than pollution. The channels overlap with those in (i) above, but more have perennial flow and some scouring.

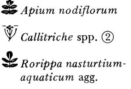

2 Larger
   blue to *purple* (with
   green)
   6+ species (4+ if *blue* or
   cover high)
   50+% cover

Common. 2—5 m wide, usually 20—30 cm deep, sometimes to *c.* 50 cm. Flow perennial, usually slow to moderate. Silty (to gravelly). Fringing herbs always emerged, usually tall, typically in (large) semi-circular clumps extending from the banks.

Often abundant: 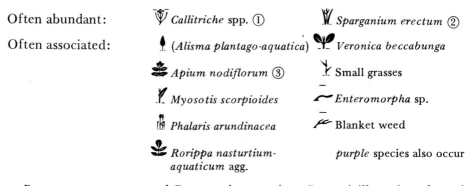 *Callitriche* spp. ①   *Sparganium erectum* ②

Often associated: (*Alisma plantago-aquatica*)   *Veronica beccabunga*

   Apium nodiflorum ③   Small grasses

   Myosotis scorpioides   *Enteromorpha* sp.

   Phalaris arundinacea   Blanket weed

   Rorippa nasturtium-aquaticum agg.   *purple* species also occur

Potamogeton natans and *Ranunculus* spp. (e.g. *R. penicillatus* long-leaved, *R. trichophyllus*) usually indicate clean streams.

Variants:
(*a*) Swift flow with coarse substrate and little vegetation
(*b*) In the west, e.g. S. Wales, where the clay is harder and equally low-lying, vegetation is closer to that of limestone, with e.g.

   Berula erecta

   Solanum dulcamara

   Veronica anagallis aquatica agg.

iii. Medium streams

purple, *purple*(—*red*)
(with *green*)
7+ species
60+% cover

Common. (3—)4—8 m wide, *c.* 30—75 cm deep, rarely more. Typically slow (to moderate) flow, with considerable silt. Fringing herbs emerged, usually tall, often in (large) semi-circular clumps growing out from the banks, particularly from gently sloping banks and during drought flows; fringing herbs also often in wispy small clumps.

Often abundant: (*Elodea canadensis*)   *Sparganium emersum* ①

   Nuphar lutea ②   Blanket weed

   (*Sagittaria sagittifolia*)

Often associated: (*Apium nodiflorum*)   *Phalaris arundinacea*

   Callitriche spp.   *Rorippa amphibia* ⑦

   Lemna minor agg. ⑤   *Rorippa nasturtium-aquaticum* agg. ③

   Myosotis scorpioides

32

Clay streams

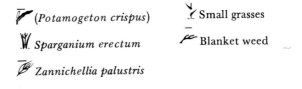

$\vphantom{}$⫟ *Schoenoplectus lacustris* ④   ⫙ Small grasses

⫙ *Sparganium erectum*   ⌒ *Enteromorpha* sp. ⑥

Ranunculus (usually long-leaved *R. penicillatus*) or *green* species usually indicate clean streams. Eutrophication increases *purple* and *red* species.

Variants:

(*a*) Undercut or vertical banks decrease emergents

(*b*) Swift flow with coarse substrates, and a depth of *c.* 30 cm leads to very little vegetation. Typical species:

⫟ (*Potamogeton crispus*)   ⫙ Small grasses

⫙ *Sparganium erectum*   ⌒ Blanket weed

⫰ *Zannichellia palustris*

(*c*) in the west, e.g. S. Wales, with harder and equally low-lying clay, vegetation is closer to that of limestone: *blue–purple.*

Often abundant: ⫚ *Berula erecta*   ⫶ *Rorippa nasturtium-aquaticum* agg.

⫚ *Callitriche* spp.   ⫙ *Sparganium erectum*

Often associated: ⫶ *Apium nodiflorum*   ⫰ *Ranunculus* e.g. *penicillatus*

⫟ *Elodea canadensis*   ⫷ Mosses

⫟ *Potamogeton crispus*   ⌒ Blanket weed

iv. Large streams

purple–red, red (with *green*)

9+ species

60+% vegetation

Very frequent. 10–20+ m wide, sides usually at least 75 cm deep, centre often deeper and without vegetation. Flow usually slow, water somewhat turbid. Silt beds frequent (or bed of clay). Fringing herbs, when present, usually wispy, always emerged.

Often abundant: ⫚ *Nuphar lutea* ①   ⫟ *Schoenoplectus lacustris* ③

⫚ *Sagittaria sagittifolia* ②   ⫟ *Sparganium emersum* ④

33

Often associated:) *Butomus umbellatus*   *Sparganium erectum* ⑤

⫪ *Elodea canadensis* ⑥   ⌒ *Enteromorpha* sp.

⟙ *Lemna minor* agg. ⑦   ⫟ Blanket weed ⑨

𝅘 *Rorippa amphibia* ⑧   and a range of *purple,*
green (and *blue*) species

Green, and a high proportion of *blue* and *purple* species usually indicate clean streams (though the converse need not apply). Eutrophication increases the proportion of *red* species.

CLAY STREAMS: C UPLAND (AND D MOUNTAIN)

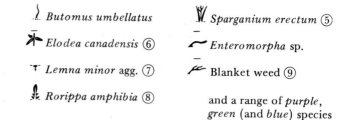

Rock: 3
Landscape: C, (D)
Streams: i, ii, iii
Colour bands: *blue,*
blue–purple, (*purple*)

Landscape with hills *c.* 800′ (345 m); the streams consequently have swifter flow. The silting, and instability of flow associated with clay mean that on these higher hills (unlike those of chalk and soft sandstone) the streams form a different habitat. There is less silt, swifter flow, and a lower trophic status than in lowland streams of equivalent size. Upland clay areas are small, and there are no large rivers.

The main upland clay areas are to the north of the Mendips and the Wessex peninsula. The main mountain areas are on the Welsh Marches.

In e.g. north-west England, streams on hard-clay (and clay-mix) have upland–mountain vegetation and are classed here.

i. Small streams without water-supported species

blue?
1+ species southern and upland, 0+ species rest
little vegetation

Rare. 1–3 m wide, shallow and sometimes summer-dry.

Probable species: *Sparganium erectum*   ⫟ Small grasses

ii + iii. Small and medium streams with water-supported species in southern and upland regions

blue, blue–purple, (*purple*)
4+ species in upper reaches, 6+ in lower ones
20+% cover

Infrequent. 2–10 m wide, *c.* 30–75 cm deep. Flow usually moderate. Substrate mixed-grained. Fringing herbs usually shorter than in the lowlands, but emerged, and typically clumped (not in bands).

Probable species:

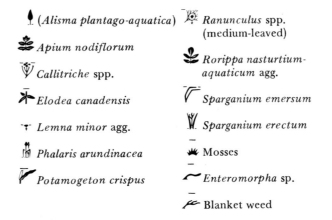

♦ (*Alisma plantago-aquatica*)	*Ranunculus* spp. (medium-leaved)
Apium nodiflorum	*Rorippa nasturtium-aquaticum* agg.
Callitriche spp.	*Sparganium emersum*
Elodea canadensis	*Sparganium erectum*
Lemna minor agg.	Mosses
Phalaris arundinacea	*Enteromorpha* sp.
Potamogeton crispus	Blanket weed

Ranunculus, *purple* and *red* species, enter in lower reaches.

Variants:
(*a*) If the flow is checked (e.g. by gates) species characteristic of lower reaches enter further upstream
(*b*) Steeper landscapes have less vegetation, approaching that listed below for northern and upland regions.

ii + iii. Small and medium streams in mountain regions and uplands in north-west England etc.

0+ species
little vegetation

Infrequent. 2—8 m wide, *c.* 20—75 cm deep. Flow usually fast (to moderate). Substrate mixed-grained (to coarse).

Probable species: Mosses

ii. Small streams with water-supported species in hard-clay lowland (e.g. north-west England)

blue (*−purple*)
3+ species
20+% cover?

Local. 1—3 m wide, 20—30 cm deep. Flow usually slow. Substrate mixed-grained.

Often abundant: *Callitriche* spp.

Often associated:

♦ *Alisma plantago-aquatica*	*Rorippa nasturtium-aquaticum* agg.
Phalaris arundinacea	*Sparganium erectum*
Potamogeton crispus	Small grasses
Potamogeton natans	

iii. Medium streams in hard-clay lowland (e.g. north-west England)

purple (with *green*)
5+ species
50+% cover

Local. 4–8 m wide, *c.* 50 cm deep. Flow slow to fast. Substrate mixed.

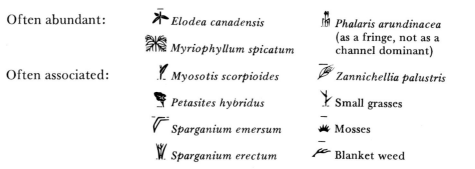

Often abundant:	*Elodea canadensis*	*Phalaris arundinacea* (as a fringe, not as a channel dominant)
	Myriophyllum spicatum	
Often associated:	*Myosotis scorpioides*	*Zannichellia palustris*
	Petasites hybridus	Small grasses
	Sparganium emersum	Mosses
	Sparganium erectum	Blanket weed

SOFT SANDSTONE STREAMS

Rock: 4
Landscape: B
Streams: i, ii, iii (–iv)
Colour bands: *blue, blue–purple*, (*purple*) with *green*

Soft sandstone country varies from nearly flat to hills occasionally reaching *c.* 800′ (345 m) high. Intermittent patches of Boulder Clay are often frequent, and the vegetation described here may perhaps be slightly influenced by these. Brook water comes from springs and run-off. In flow, bed stability, silting, and water clarity, sandstone streams are intermediate between chalk and clay ones. Sand is common on the beds, and in swifter reaches may cause unstable substrates and sparse vegetation. The substrate is mesotrophic, but differs from that of the equally mesotrophic chalk in having more silt on the beds, higher concentrations of nutrients (particularly calcium, nitrogen, sodium and phosphate-phosphorus) in the silt and a lower hardness ratio.

Soft sandstone mainly outcrops in southern England, the Cheshire–Shropshire region, and East Anglia. Large streams are few, and mainly too polluted to determine their undamaged vegetation.

i. Small streams without water-supported species

blue
2+ species
20+% cover

Soft sandstone streams

Infrequent. 1–3 m wide, usually dry for part or all of the summer. If wet, then shallow and flow usually slow. Substrate may be sandy.

Probable species: 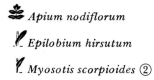 *Apium nodiflorum*   *Sparganium erectum* ①

Epilobium hirsutum   *Veronica beccabunga*

Myosotis scorpioides ②   Small grasses

Variant:
(*a*) Grade into (sunk) ditches

ii. Small streams with water-supported species

blue (*blue–purple*, with *green*)
3+ species
30+% cover

Frequent. 1–3 m wide, usually not over 30 cm deep. Usually with moderate or slow perennial flow and a mixed-grained substrate. Fringing herbs are short, as in chalk streams, but are emerged, not forming submerged carpets, and more often in clumps than bands.

Often abundant:  *Apium nodiflorum*   (Blanket weed)

Callitriche spp.

Often associated: *Lemna minor* agg.   *Sparganium erectum*

Myosotis scorpioides   *Veronica beccabunga*

Rorippa nasturtium-aquaticum agg.   Small grasses

Variants:
(*a*) In slower, more silty flows *Potamogeton natans* and more eutrophic species may occur
(*b*) Smaller streams, particularly if nearly or quite dry for part of the summer, have 2+ species, probably:

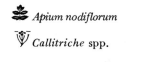 *Apium nodiflorum*   *Rorippa nasturtium-aquaticum* agg.

Callitriche spp.   *Sparganium erectum*

(*c*) Steep or undercut banks may sharply decrease fringing herbs

iii(–iv). Medium (and to large) streams

blue to *purple*, with *green*
6+ species (5+ if without *purple*)
60+% cover

37

Frequent. 4–8 m wide, *c.* 20–75 cm deep, sometimes deeper. Flow slow and moderate. Fringing herbs are short, as in chalk streams, but are emerged, not forming submerged carpets, and more often in clumps than bands.

Often abundant:

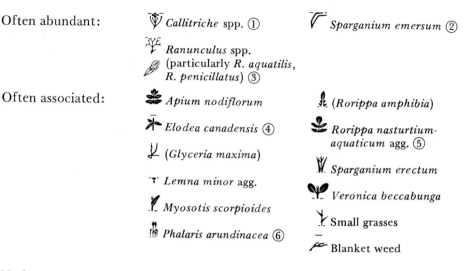

Callitriche spp. ①   *Sparganium emersum* ②

Ranunculus spp. (particularly *R. aquatilis*, *R. penicillatus*) ③

Often associated:

Apium nodiflorum   (*Rorippa amphibia*)

Elodea canadensis ④   *Rorippa nasturtium-aquaticum* agg. ⑤

(*Glyceria maxima*)   *Sparganium erectum*

Lemna minor agg.   *Veronica beccabunga*

Myosotis scorpioides   Small grasses

Phalaris arundinacea ⑥   Blanket weed

Variants:
(*a*) Slow streams are without *Ranunculus* and may have, or be dominated by, *Potamogeton natans*
(*b*) Streams in less fertile regions may have vegetation approaching that on New Forest acid sands (see below)
(*c*) Undercut or very steep banks decrease emergents

NEW FOREST SANDS STREAMS

Rock: 5
Landscape: B
Streams: ii, iii(–iv)
Colour bands: *orange* to *purple* (without *yellow*) with *green*

These streams rise on very infertile land, but downstream eutrophication is unusually great as the silt from the sandstone accumulates in the lower reaches; the change from *orange* to *purple* can be rapid. Intermediates occur, of course, between the nutrient-poor and nutrient-rich variants described below. See also the description of soft sandstone streams above. The streams are found in and near the New Forest, Hampshire.

ii. Small streams with water-supported species

1 *orange* (with *green*)
  3+ species
  50+% cover
Local. 1–3 m wide, shallow, may dry in summer. Sandy.

New Forest sands streams

Probable species: 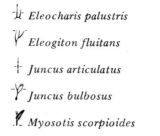 *Eleocharis palustris*

Eleogiton fluitans

Juncus articulatus

Juncus bulbosus

Myosotis scorpioides

Myriophyllum alterniflorum
(wetter)

Oenanthe crocata

Potamogeton natans
(wetter)

Potamogeton polygonifolius

Small grasses

2 *blue*
  1+ species
  10+% cover
Infrequent. 1–3 m wide, 20–40 cm deep. With perennial flow and humus-stained water. Mostly shaded, with *Callitriche* spp. the most frequent in such streams.

Probable species: *Callitriche* spp.

Catabrosa aquatica

Ranunculus aquatilis

Ranunculus omiophyllus

Sparganium emersum

Veronica beccabunga

iii(–iv). Medium streams

1 *blue*, with *green*
  5+ species
  60+% cover
Local. 4–8 m wide, mostly less than 50 cm deep. Usually with moderate flow and humus-stained water. Sandy.

Probable species: *Apium nodiflorum*

Callitriche hamulata

Callitriche spp.

Mentha aquatica

Oenanthe crocata

Phalaris arundinacea

Potamogeton natans

Ranunculus peltatus

Rorippa nasturtium-aquaticum agg.

Sparganium emersum

Sparganium erectum

2 *purple*
  7+ species
  60+% cover
Local, downstream of (1). 5–10+ m wide, mostly 50 cm–1 m deep. With slow flow, and somewhat turbid water. Silty.

Probable species: *Alisma plantago-aquatica*

Callitriche spp.

Elodea canadensis

Lemna minor agg.

Myosotis scorpioides

Nuphar lutea

Phalaris arundinacea

Sparganium emersum

Sparganium erectum

Blanket weed

3 Descriptions of stream communities

ALLUVIAL STREAMS

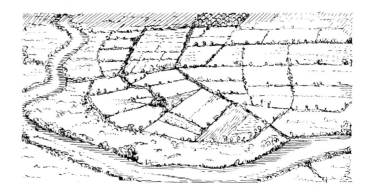

Rock: 6
Landscape: A
Streams: (ii), iii, (iv)
Colour bands:
turquoise/blue—red

These streams rise on the higher ground, usually clay, outside the plain, and when in the plain the vegetation gradually changes from that of clay (see above) to being somewhat like that of dykes and drains (see Chapter 5). As the watercourses still carry silt and water from the high ground, particularly during storm flows, the vegetation does not become that of an undamaged drain (when on silt or peat); because of the extra silting, it is more eutrophic. As the stream leaves the higher ground the flow becomes slow, the channel dyke-like in section (see figure on p. 98) and the substrate fine-grained (silt on a silt subsoil or peaty silt on a peat one). The management is that of drains (see Chapters 5 and 6) and the larger streams, particularly in the Fenland, are much-managed, bearing only vegetation tolerant of this.

For the list of *turquoise* species see the dykes and drains chart (Table 5.1, p. 100).

These streams are necessarily confined to the larger alluvial plains (see rock type map, pp. 126–45).

Streams largely on clay, even in the alluvial plain

Also see the information on clay dykes in Chapter 5 (p. 98).

Streams largely on peat or silt. ii + iii. Smaller channels

turquoise/blue—red
8+ species
80+% cover

Local. (3–)4–8(–10) m wide, usually at least 50 cm deep. Little water movement except in storms.

Probable species: ※ *Ceratophyllum*   🌿 *Nuphar lutea* ⑤
        demersum ①
        🌾 *Phalaris arundinacea* ⑦
        ⊤ *Lemna minor* agg. ③
        🔺 *Sgaittaria sagittifolia*

Sparganium emersum ②   *Enteromorpha* sp. ⑥

Sparganium erectum ④   Blanket weed

Variants:

(*a*) Further from the high ground there is an increase in species which are in the dykes and drains chart (Table 5.1, p. 100) and not on the stream dial

(*b*) The more the peat, the higher the proportion of *turquoise/blue* species

(*c*) The more the silt, the higher the proportion of *red* species (though a *red* colour band indicates damage)

CARE. Side channels to alluvial streams, sited solely in the plains, count as dykes, not streams, and should be looked up in Chapter 5.

STREAMS ON MIXED CATCHMENTS (SOFT ROCKS)

Rocks: 1, 2, 3, 4
Landscape: B
Streams: (i, ii), iii, iv
Colour bands: *blue* to *purple—red*, with *green*

A mixed catchment is one which, upstream of the site recorded, consists of two or more rock types. Vegetation usually reflects the proportions of each, subject to the following rules:

1. The rock types of the headwaters have the most influence on the vegetation, and those near the mouth the least.
2. When the proportions of the rock types are equal, that at the site is the more important.
3. Limestone and clay have a stronger influence than sandstone.
4. In a mixed mesotrophic—eutrophic habitat, deep slow flows, allowing silting, lead to a *redder* vegetation, and swifter shallower ones with coarser substrates to a *bluer* one. This change can occur both as a pattern within one reach of a river and as a difference between summers (low flows in larger rivers producing shallower, more turbulent and thus more eroding flows and *bluer* vegetation).
5. The change in plant community after crossing a rock boundary occurs much sooner if tributaries enter which rise on the new rock type.
6. The rock type map (pp. 126—45) is only on a small scale. River plants are very sensitive to rock type, and when vegetation (*a*) does not fit that expected from the rock type shown on the map and (*b*) is species-rich and does not show a characteristic damage pattern (see Chapter 4), a large-scale geological map should be consulted. It will probably be found that the site is on a small patch of, say, limestone in a clay area; or thick Boulder Clay may, locally, consist of sand instead of clay. (Some adjustments have been made in the rock type map: e.g. lowland Coal Measures in an otherwise clay catchment are, vegetationally, clay, and very fertile sandstones also bear clay vegetation.)

Necessarily, most streams flowing on two rock types are quite large. The few small ones occur either where there is a mosaic of rock types (e.g. parts of the Northamptonshire Wolds) or where Boulder Clay is thick enough to influence the river plants but not so thick as to obscure the effect of the solid rock below (e.g. some tributaries of R. Stour, Essex).

3 Descriptions of stream communities

SOFT LIMESTONE–CLAY STREAMS

Rocks: 1, 2, 3
Landscape: B
Streams: (i, ii), iii, iv
Colour bands: *blue* to *purple* (with *green*)

Derive the expected vegetation from that of clay, and of chalk or oolite (as appropriate), in accordance with rules 1 to 6 above and the following points:
1. In streams of size (iii) – and also of sizes (ii) and (iv) – species occurring more frequently on mixed catchments than on clay or limestone alone are:

Groenlandia densa   *(Potamogeton pectinatus?)*

Myriophyllum spicatum   *Potamogeton perfoliatus*

Oenanthe fluviatilis   *Zannichellia palustris*

Potamogeton crispus

2. In streams of sizes (iii) and (iv) – and rarely of size (ii) – when considering catchments with different proportions of the two rocks, the main increase on passing from full limestone to full clay, of

Nuphar lutea   *Sparganium emersum*

occurs when the clay influence first starts, when there is little clay in the catchment.
3. Similarly, the main increase in all sizes of streams on passing from full limestone to full clay, of

Sparganium erectum

is when the catchment is half limestone, half clay.
4. The *Ranunculus* species present also change with increasing proportions of clay in the catchment. *Ranunculus calcareus* and medium-leaved *Ranunculus penicillatus* usually disappear before the catchment is half clay. At that stage, *Ranunculus fluitans*, long-leaved *Ranunculus penicillatus* and *Ranunculus trichophyllus* are likely. *Ranunculus fluitans* is usually lost when the catchment is mainly clay, while the other two persist but increasingly sparsely.
5. Boulder Clay may lend instability to the bed of a limestone stream, and decrease the vegetation.

SOFT LIMESTONE–SANDSTONE STREAMS

Rocks: 1, 2, 4
Landscape: B
Streams: (i, ii), iii, iv
Colour bands: *blue* to *blue–purple*

Derive the expected vegetation from that of sandstone and of chalk or oolite (as appropriate), in accordance with rules 1 to 6 above and the following points:
1. Streams rising in fertile sandstone but with most of their length on chalk have some eutrophic influence (*purple* or *red* species) throughout.
2. Small proportions of infertile sandstone in the catchment may have no influence on the limestone reaches.

CLAY–SOFT SANDSTONE STREAMS

Rocks: 3, 4
Landscape: B
Streams: (i, ii), iii, iv
Colour bands: *blue* to *purple–red*

Derive the expected vegetation from that of clay and that of soft sandstone, in accordance with rules 1 to 6 above.

RESISTANT ROCK STREAMS

Rock: 7
Landscapes: B, C, D, E, F, G, H
Streams: i, ii, iii, v
Colour bands: *brown* to near-*purple*

Resistant rock landscapes vary from almost flat to steeply mountainous. Brook water comes from run-off which, in the hillier regions with high rainfalls, leads to unstable flows of great water force. The streams vary greatly, and many intermediates occur between the vegetation types listed here. The descriptions of several possible types should be consulted if the identification is doubtful. Because of the variation with small changes in topography etc., damage is difficult to diagnose (see Chapter 4), and can be deduced only when diversity, cover and colour band differ from those given for all possible options or when the directions given in Chapter 4 apply. The topographical type of a stream depends on the water force it receives from run-off etc., and is determined primarily by the hilliest topography in the catchment and secondarily by the proportion of the stream to be found in each landscape type. Where acid peat (from the land around, or from the subsoil) occurs on the channel bed, the streams are the most nutrient-poor of any found in Britain (they are found with landscape H, less with G, and less again with F). The inorganic silt (found in B, C, D, E and, with peat as well, in F and G) has the lowest nutrient status in Britain, being particularly low in calcium, phosphate-phosphorus, chloride and sulphate-sulphur and the hardness ratio is as low as on soft sandstone. In swift streams, with stony and bouldery substrates and very little silt, nutrient status is again very low, though higher than in streams where acid peat occurs on the bed.

Fringing herbs in shallow streams without excessive water force are short, emerged, well-anchored, and more in clumps than bands. Emerged carpets can occur where several species are co-dominant. In larger streams the plants become more wispy and are confined to the sides.

Resistant rock streams occur over most of highland Britain, which covers the north and west, from Sutherland to Cornwall.

RESISTANT ROCK STREAMS: H BLANKET BOG (= BOG PLAINS)

Rock: 7
Landscape: H
Streams: i, ii
Colour bands: *brown*
(*brown–orange*)

Bog streams occur in blanket bogs in almost flat watersheds, particularly in the Scottish Highlands. The substrate is acid peat (e.g. *Sphagnum*). Water movement is little, except after rain, so water force is low and fierce spate flows are absent. The water is peat-stained.

They are found mostly in the Scottish Highlands, but also occur in wet peaty places elsewhere.

i. Small streams without water-supported species

brown
1+ species
25+% cover

Rare, but may be locally frequent. 0.5—3 m wide, often *c.* 10 cm deep.

Probable species: (*Carex* spp.) (*Menyanthes trifoliata*)

Eriophorum angustifolium ① (*Sphagnum* spp.)

(*Juncus* spp.)

ii. Small streams with water-supported species

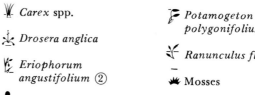

brown (brown—orange)
4+ species
40+% cover

Rare, but may be locally frequent. *c.* 2—4 m wide, often 20—30 cm deep.

Probable species: *Carex* spp. *Potamogeton polygonifolius* ①

Drosera anglica *Ranunculus flammula*

Eriophorum angustifolium ② Mosses

Menyanthes trifoliata ③

Note. In this habitat, count the mosses as *brown*, not *yellow*.

RESISTANT ROCK STREAMS: G RIVERS WITH MUCH BLANKET BOG

Rock: 7
Landscape: G
Streams: iii, v
Colour bands:
brown to *orange*

The streams have fine peaty soil in sheltered parts (edges etc.), and vegetation is almost confined to these peaty places. The main flow is usually swift, the spates are of considerable force, and the substrate is normally coarse, often bouldery. The water is peat-stained and of variable depth.

Landscape types with much blanket bog have been subdivided into categories H, for small streams entirely on blanket bog, and G, for larger ones flowing in catchments with much blanket bog but having hills as well. The division is because streams of sizes (iii) and (v) do not necessarily arise as ones of sizes (i) or (ii), but may start in the hills, i.e. on landscapes D or E (rarely C or F).

The habitat depends on the joint effect of (*a*) the amount of erodable blanket bog and similar acid peat in the catchment, which determines the amount of peat being washed into the stream, and (*b*) the flow regime, as over-swift flows prevent any of this peat being deposited on the stream bed. As there are many possible variations in these two factors, the absence or impoverishment of the community is more likely to be because the flow is too swift or the bog too small than because the site is damaged. Note that the division between stream sizes (iii) and (v) is here given on vegetation rather than on width: i.e. on a flow (plus substrate) character.

These streams occur mainly in the north of the Scottish Highlands.

iii. Medium streams

brown, brown—orange
4+ species
5+% cover

Local. *c.* 4—14 m wide, usually *c.* 50—75 cm deep.

Probable species:

⚘ *Carex* spp.	⚘ *Littorella uniflora* ①
⚘ *Eleocharis acicularis*	⚘ *Potamogeton polygonifolius* ②
⚘ *Juncus articulatus* ③	⚘ *Ranunculus flammula* ④

v. Large spatey streams

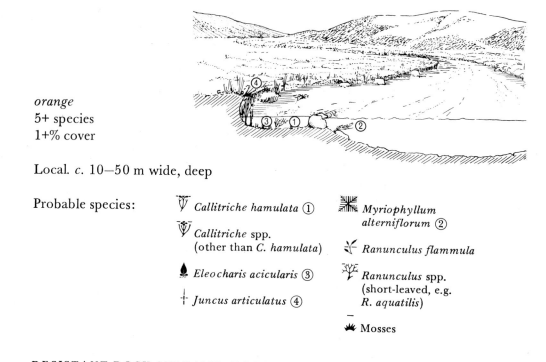

orange
5+ species
1+% cover

Local. *c.* 10–50 m wide, deep

Probable species:

Callitriche hamulata ①

Callitriche spp.
(other than *C. hamulata*)

Eleocharis acicularis ③

⊤ Juncus articulatus ④

Myriophyllum
alterniflorum ②

Ranunculus flammula

Ranunculus spp.
(short-leaved, e.g.
R. aquatilis)

Mosses

RESISTANT ROCK STREAMS: F LOWLAND MOORLAND

Rock: 7
Landscape: F
Streams: (i), ii, iii, (iv/v)
Colour bands: *orange*
to *blue*

These streams occur in nearly flat, peaty areas, where the peat entering the streams is less, and more mineralised than in landscape types G and H above. The spate force is low, flow usually moderate to fast, substrate mixed-grained to rather coarse (but rarely bouldery), and the water is hardly peat-stained. There are few large rivers, since if the streams are long enough to become large they usually flow into steeper topography and are described under C or D below. Large rivers cannot be assessed for damage since, as in G above, the absence or impoverishment of the community is likely to be because the flow, or spate force, has become greater. Even smaller streams can be difficult to assess, as a slight variation in the fertility of the land will, by bringing more inorganic silt, or more acid peat, into the stream, move the colour band towards *blue* or *orange* respectively. Accurate diagnoses can be made from changes in colour band, cover or diversity over a period of years.

The two largest areas in which these streams are found are in south-west England (Bodmin Moor, Dartmoor, etc.) and in the Solway peninsula.

i. Small streams without water-supported species

orange to *blue*
1+ species
5+% cover

46

Resistant rock streams

Rare. *c.* 0.5—2 m wide, often dry in late summer. Often shaded.

Probable species: 🌿 *Apium nodiflorum*   🌿 *Oenanthe crocata*

                          Myosotis scorpioides   *Ranunculus omiophyllus*

ii. Small streams with water-supported species

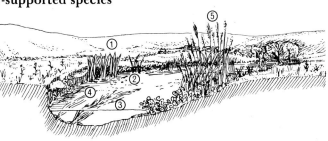

orange, *yellow*, with *green*
5+ species
(2+ if less than 2 m wide)
40+% cover

Infrequent. *c.* 0.5—4 m wide, usually 20—50 cm deep.

Probable species
widespread: 🌱 *Callitriche* spp.   🌱 *Sparganium erectum* ①

                      Potamogeton natans ②

if northern: 🌑 *Eleocharis acicularis*   *Phalaris arundinacea*

                 † *Juncus articulatus*   *Sparganium emersum*

                 Mentha aquatica

if southern: *Eleogiton fluitans* ③   *Myosotis scorpioides*

                 Glyceria fluitans ④   *Phragmites communis* ⑤
                 (with long floating leaves)
                                         Ranunculus omiophyllus
                 Juncus bulbosus

Variants:
(*a*) Narrow channels with little water in summer are likely to bear *Callitriche* spp. with one or two species from (i) above
(*b*) With more acid peat, *Littorella uniflora* and perhaps other *brown* species will be present

iii. Medium streams

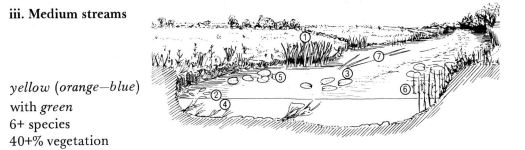

yellow (*orange—blue*)
with *green*
6+ species
40+% vegetation

Not common. 4—10 m wide, usually 30—75 cm deep.

Probable species widespread:		
	Callitriche spp.	*Sparganium erectum* ①
	Potamogeton natans ②	Mosses

if northern:		
	Callitriche hamulata	*Nuphar lutea* ③
	Eleocharis acicularis ④	*Nymphaea alba* ⑤
	Equisetum palustris ⑥	*Sparganium emersum* ⑦

if southern:		
	Eleogiton fluitans	*Juncus bulbosus*
	Glyceria fluitans (with long floating leaves)	*Myosotis scorpioides*
		Oenanthe crocata

Variants:

(*a*) Swifter streams with coarser substrates are species-poor with vegetation confined to local sheltered areas

(*b*) With more acid peat, *Littorella uniflora* and perhaps other *brown* species will be present

iv/v. Large streams

orange—yellow
3+ species
5+% vegetation

Rare. At least 10 m wide, often deep. Vegetation sparse and confined to local sheltered areas. Species as in (iii) above.

RESISTANT ROCK STREAMS: B LOWLAND FARMLAND

Rock: 7
Landscape: B
Streams: i, ii, (iii), iv/v
Colour bands: *yellow* to *purple*, with *green*

Two categories of streams are placed here: small streams rising in the lowlands (either in lowland areas, as in parts of Cornwall, or on the fringes of hills), and streams rising in the hills but flowing into lowlands or small flood plains for long enough for their vegetation to be affected by this flatter ground. The latter, though rising on Resistant rock, may be on this, alluvium or soft rocks in their lowland reaches but, for convenience, all are grouped here.

The soil is more fertile than in landscape types F, G or H, but it varies with the details of landscape and land use, and the vegetation is usually sparse. Narrow streams are often shaded or with eroding flows, and silty flood plain rivers have this silty substrate disturbed during spates. The water is not peat-stained.

Lowland farmland can be arable and fertile, even where tributaries arise. Most rivers of this type are in north-east Scotland (particularly the Aberdeenshire area), where the upper tributaries rise in the uplands; their vegetation is described under 'C upland' below (p. 51).

It is also possible to have slower-flowing minor tributaries in alluvial valleys, flood plains, etc. Their vegetation varies with the local habitat factors, but can be related to types described in this book (and see *River Plants* for habitat preferences of species, and how to deduce habitat factors from these).

STREAMS RISING IN THE LOWLANDS

i. Small streams without water-supported species

yellow to *purple*
1+ species
1+% vegetation

Local. 1—4 m wide, shallow. Usually slow to moderate flow. Fine- to coarse-grained substrates.

Probable species:

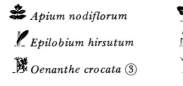

Apium nodiflorum

Epilobium hirsutum

Oenanthe crocata ③

Petasites hybridus ①

Phalaris arundinacea ②

Grasses

Variants:
(*a*) Grading at one extreme almost to ditches with tall emergents dominant
(*b*) At the other extreme there are almost-empty hill streams (see 'C upland' and 'D mountain': pp. 51 and 56)

ii + (iii). Small (to medium) streams with water-supported species

yellow to *blue*
2+ species
5+% cover

Local. *c.* 1—5 m wide, *c.* 20—40 cm deep. Usually moderate flow. Mixed to coarse substrates. The wider streams come from the less fertile catchments.

Probable species:

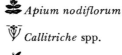
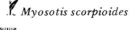
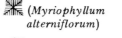

Apium nodiflorum

Callitriche spp.

Myosotis scorpioides

(*Myriophyllum alterniflorum*)

Oenanthe crocata

Ranunculus
(short-leaved, e.g.
R. aquatilis)

Rorippa nasturtium-aquaticum agg.

Veronica beccabunga

Mosses

iii. Medium streams

blue to *blue—purple*
4+ species
10+% cover

Rare. *c.* 4—8 m wide. With usually moderate flow and mixed-grain substrates. (See the section on small to medium streams above for streams of this width on the less fertile catchments.)

Probable species:

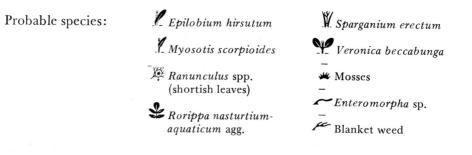

Epilobium hirsutum

Myosotis scorpioides

Ranunculus spp. (shortish leaves)

Rorippa nasturtium-aquaticum agg.

Sparganium erectum

Veronica beccabunga

Mosses

Enteromorpha sp.

Blanket weed

STREAMS RISING IN THE HILLS

i + ii + (iii). Small (to medium) streams

yellow—blue to *blue—purple*

The vegetation of these streams varies greatly with the steepness of the hills on which they rise, the comparative length of the stretches on hills and on flatter ground, and the nature of the flatter ground (more silt, and therefore e.g. *Glyceria maxima* in a small alluvial plain, more mixed-grained substrates, and therefore e.g. *Elodea* (slower) and *Ranunculus* spp. (faster) on Resistant rock). Consult the descriptions of streams rising in the lowlands; also the sections on upland and mountain below (pp. 51 and 56), those on streams in alluvial plains above (p. 40) and the section in Chapter 6 on altering flow types (p. 116).

iv/v. Large streams

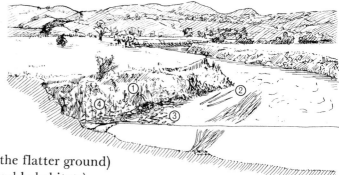

green to *purple*
4+ species (when far into the flatter ground)
5+% cover (50+% in favourable habitats)

Local. 8—20+ wide, deep. With slow flow and with some to much silt.

When a hill river first enters a flood plain it keeps the vegetation of the hill river. In the longer flood plains or lowlands, the vegetation changes slowly. The species of faster flow listed below occur more towards the hills, and those of slower flow towards the mouth. In very long flood plains on fertile rocks (e.g.

rivers flowing from the Pennines to the Plain of York), the vegetation may approach that of clay streams.

Probable species:

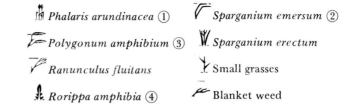

Phalaris arundinacea ①   *Sparganium emersum* ②

Polygonum amphibium ③   *Sparganium erectum*

Ranunculus fluitans   Small grasses

Rorippa amphibia ④   Blanket weed

RESISTANT ROCK STREAMS: C UPLAND

Rock: 7
Landscape: C
Streams: i, ii, iii, v
Colour bands: *yellow*
to *blue—purple*

The land use in upland areas is mostly grassland, with arable (and orchards, etc.) below, and perhaps more rough pasture on the hilltops. Upland streams may also occur within mountainous regions, when streams are so sited as to receive little run-off. The reverse also applies: mountain streams can occur in upland regions. Spates are frequent, but their force is low (though on the rare occasions when a really fierce spate does occur there is long-term damage to the vegetation). Flow is fairly swift, and there are frequently alternating stretches of swift shallow coarse channels, and slow deeper siltier ones. The substrates are usually basically mixed-grained and often coarse (but hardly ever bouldery). (Swifter normal flows and fiercer spates convert upland to mountain rivers.)

As mentioned in the section on lowland farmland above, there are also fertile, mainly arable regions, with richer vegetation — particularly in north-east Scotland — where the upper tributaries rise in uplands but much of the catchment is low-land. Stream vegetation is often similar to that of hard sandstone in a rather steeper landscape. Also placed in this category are the lower reaches of larger rivers which rose in the mountains and have flowed through flatter landscapes (not flood plains) for many miles.

USUAL TYPE

i. Small streams without water-supported species

1 Narrow (*c.* 0.5—1 m) summer-dry channels on wide (flat) hilltops with rough
  pasture. Locally very frequent. Often 100% cover.
The characteristic, and usually the only species:

Juncus effusus ①

Grazed (grassy) channels also occur.

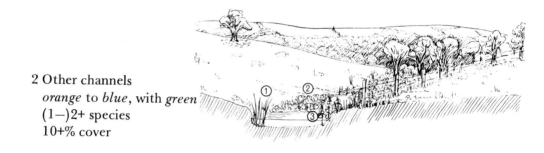

2 Other channels
  orange to *blue*, with *green*
  (1—)2+ species
  10+% cover

Infrequent. 1—4 m wide, usually 10—30 cm deep. Perennial eroding flows.

Probable species: *Apium nodiflorum*   *Petasites hybridus*

  Juncus articulatus ①   *Rorippa nasturtium-
              aquaticum* agg. ②
  Mentha aquatica ③

  Oenanthe crocata   *Veronica beccabunga*

      Small grasses

Variant:
(*a*) Channels with low run-off at the higher altitudes may have emerged short
   carpets of fringing herbs, colour band *blue*, with 4+ species and 60+% vegetation, with much silt

ii. Small streams with water-supported species

blue
3+ species
20+% cover

Resistant rock streams

Frequent. 1—4 m wide, usually 10—30 cm deep. Usually with eroding flows.

Probable species:

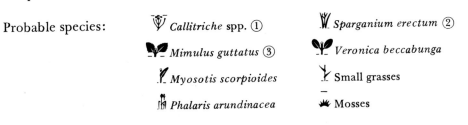

Callitriche spp. ①

Sparganium erectum ②

Mimulus guttatus ③

Veronica beccabunga

Myosotis scorpioides

Small grasses

Phalaris arundinacea

Mosses

Variant:
(*a*) As in (i2) above, but also with *Callitriche* spp. (or mosses)

iii. Medium streams

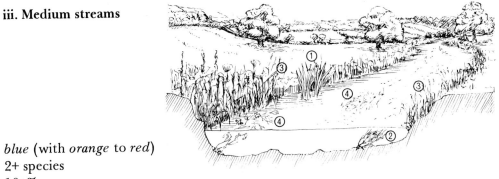

blue (with *orange* to *red*)
2+ species
10+% cover

Frequent. 4—8(—10) m wide, usually 30—60 cm deep. Wash-out during storm flows more likely than in (i) and (ii).

Probable species:

Callitriche spp.

Sparganium erectum ①

Elodea canadensis ②

Veronica beccabunga

Mimulus guttatus
(northern)

Mosses

Myosotis scorpioides

Blanket weed

Phalaris arundinacea ③

Ranunculus spp.
(particularly short-leaved
R. aquatilis, *R. peltatus*
and *R. penicillatus*) ④

v. Large streams

blue
1+ species (may be much higher)
5+% cover (may be much higher)

Infrequent. 8—20+ m wide, usually 0.5 m or more deep. Wash-out in storms more damaging than in (iii).

3 Descriptions of stream communities

Probable species:    *Apium nodiflorum*

       Callitriche spp.

       (Epilobium hirsutum)

       (Myosotis scorpioides)

       (Petasites hybridus)

       Phalaris arundinacea

       Ranunculus spp.
(short to medium-leaved,
particular *R. aquatilis*,
R. peltatus, *R. penicillatus*)

       Sparganium erectum

       Blanket weed

Variant:

(*a*) With decreased spates (e.g. from reservoir sited upstream), either: greater diversity; or high cover of *Ranunculus* spp., with or without high diversity

FERTILE LANDSCAPE, NORTH-EAST SCOTLAND, ETC.

i. Small streams without water-supported species

blue(*–yellow*)
2+ species
20+% cover, often 75+%

Locally frequent. 0.5–3 m wide, usually 10–30 cm deep. Moderate to fast flow. Mixed substrate.

Often abundant: *Mimulus guttatus*

       Phalaris arundinacea

       Rorippa nasturtium-aquaticum agg.

Often associated: *Myosotis scorpioides*   Small grasses

Variants:

(*a*) More oligotrophic upper tributaries, with the above vegetation and one or more of:

       Caltha palustris   *Petasites hybridus*

       (Mentha aquatica)

(*b*) Dry grassy channels

(*c*) Swifter streams with sparse grass or other vegetation

ii. Small streams with water-supported species

blue(*–yellow*)
4+ species
30+% cover (less with higher water force)

Locally frequent. 1–3 m wide, usually 10–30 cm deep. Typically have alternating zones of fast and slow clear water. Stony or mixed substrates. Water sometimes peat-stained.

Often abundant: *Callitriche* spp.   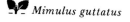 *Mimulus guttatus*

       (Callitriche hamulata)

54

Resistant rock streams

Often associated:

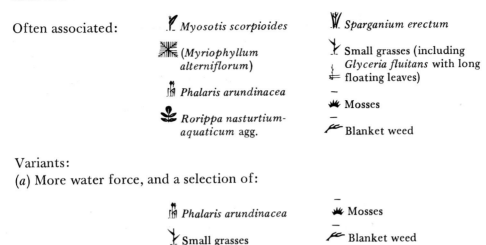

Myosotis scorpioides

(Myriophyllum alterniflorum)

Phalaris arundinacea

Rorippa nasturtium-aquaticum agg.

Sparganium erectum

Small grasses (including Glyceria fluitans with long floating leaves)

Mosses

Blanket weed

Variants:

(a) More water force, and a selection of:

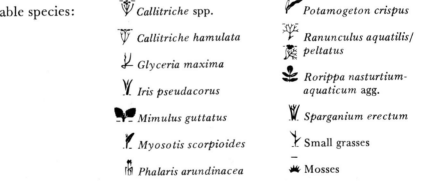

Phalaris arundinacea

Small grasses

Mosses

Blanket weed

(b) Channels with much-grazed banks have more fringing herbs

(c) Channels with ungrazed banks have fewer fringing herbs and more *Phalaris arundinacea*

iii. Medium streams

blue (with *orange* to *purple*)
6+ species
20+% cover

Local. 4—8 m wide, usually 30—75 cm deep. Alternating zones of moderate and faster or slower flow. Mixed or (firm) stony substrates. Water often peat-stained.

Probable species:

Callitriche spp.

Callitriche hamulata

Glyceria maxima

Iris pseudacorus

Mimulus guttatus

Myosotis scorpioides

Phalaris arundinacea

Potamogeton crispus

Ranunculus aquatilis/peltatus

Rorippa nasturtium-aquaticum agg.

Sparganium erectum

Small grasses

Mosses

(other *orange* and *yellow* species may also occur)

iv/v. Large streams

blue—purple (with *green*)
6+ species
30+% cover

Local. 10+ m wide, 0.5+ m and deep. Flow slow or intermittently fast. Substrate mixed-grained.

Often abundant:

Ranunculus aquatilis/peltatus

Often associated:

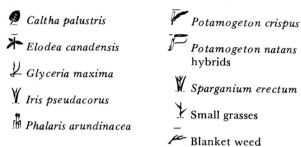

🍃 *Caltha palustris*	*Potamogeton crispus*
Elodea canadensis	*Potamogeton natans* hybrids
Glyceria maxima	*Sparganium erectum*
Iris pseudacorus	Small grasses
Phalaris arundinacea	Blanket weed

RESISTANT ROCK STREAMS: D MOUNTAIN

Rock: 7
Landscape: D
Streams: i, ii, iii, v
Colour bands: *blue* to *blue—purple* (but species from *orange* to *red*, with *green*)

Mountain landscapes comprise the greater part of the Resistant rock areas. Land use is predominantly grassland below, and mountain and moorland vegetation on the hilltops. Locally, upland or very mountainous streams may occur in mountain regions, where local topographical variations lead to unusually low, or unusually high, water force respectively, and similarly mountain streams can occur in upland regions (with unduly high water force for the area) or very mountainous ones (with unduly low water force). Fierce spates are frequent, and this and the characteristically swift normal flow usually prevent much bulk of vegetation developing except in wide and long lower reaches where water force has diminished. Alternating reaches of swifter coarser shallower channels, and of slower finer-particled deeper ones, are common. The swifter reaches frequently have rapid flow, with the water broken by boulders or large stones, and the slow reaches are only rarely silty (more often gravelly or stony).

i. Small streams without water-supported species

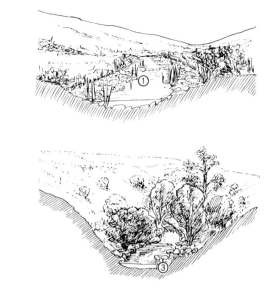

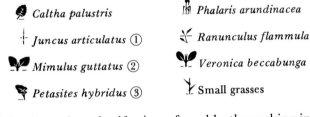

blue (green, orange)
0+, 2+ species
0+% cover

Frequent. 1–4 m wide, *c.* 10–40 cm deep. Perennial eroding flow. Where flow is less fierce, fine particles are deposited at the sides, and plants can grow well there.

Probable species:

Caltha palustris

Juncus articulatus ①

Mimulus guttatus ②

Petasites hybridus ③

Phalaris arundinacea

Ranunculus flammula

Veronica beccabunga

Small grasses

The small grasses (mainly *Agrostis stolonifera*) are found both reaching into the water from the banks and as temporary patches in the centre of the channel that have been washed down from upstream.

Variants:
(*a*) Fringing *Juncus articulatus* is common, and frequently the sole species, in streams with gently sloping banks on areas of open peaty moorland
(*b*) *Petasites hybridus* is often found in rather shaded streams with medium-sloping banks
(*c*) Dry channels with *Juncus effusus* (and sometimes other species)
(*d*) Damp grassy channels

ii. Small streams with water-supported species

yellow to *blue*, with *green*.
1+ species
1+% cover

Common. 2–4 m wide, *c.* 20–50 cm deep. Eroding flow but less scour in the channel centre than is typical of (i).

Probable species: 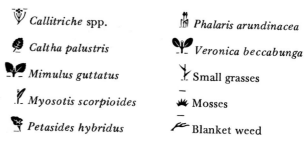   *Callitriche* spp.   *Phalaris arundinacea*

                            Caltha palustris   *Veronica beccabunga*

                            Mimulus guttatus   Small grasses

                            Myosotis scorpioides   Mosses

                            Petasides hybridus   Blanket weed

The small grasses found are as in (i).

Variants:
(*a*) *Juncus articulatus* type as in (i), but also with mosses (or *Callitriche*)
(*b*) *Petasites hybridus* type as in (i)
(*c*) *Orange* species in more peaty landscapes

iii. Medium streams

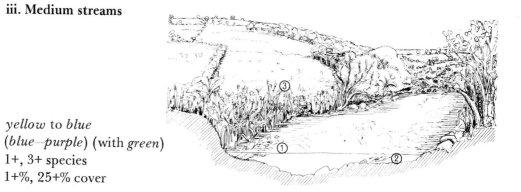

yellow to *blue*
(*blue—purple*) (with *green*)
1+, 3+ species
1+%, 25+% cover

Common. 4—8 m wide, *c.* 30 cm—1 m deep, often varying within one reach. May have 75% cover in some flood plain areas.

Probable species: *Callitriche* spp.   *Ranunculus* spp.
                                                (short-leaved, mainly

                        Juncus articulatus   *R. aquatilis*) ①

                        Mimulus guttatus   *Sparganium erectum*

                        Myriophyllum   Small grasses

                           alterniflorum ②   Mosses

                        Petasites hybridus   Blanket weed

                        Phalaris arundinacea ③

Variants:
(*a*) In more peaty landscapes:

                         Callitriche hamulata   *Juncus articulatus*

                        Eleocharis palustris   *Oenanthe crocata*

(*b*) With more inorganic silt and shelter:

                        Elodea canadensis   *Potamogeton crispus*

                        Myriophyllum spicatum

(*c*) In parts of central Scotland, important species are
Potamogeton natans hybrids, e.g. *Potamogeton* × *sparganifolius*

v. Large rivers

yellow to *blue—purple*, with *green*
0+, 4+ species
0+%, 25+% cover

Common. *c*. 8—30+ m wide, *c*. 0.5—2 m deep usually, often varying within one reach.

Probable species:

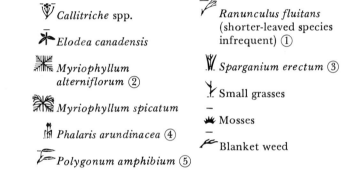

Callitriche spp.

Elodea canadensis

Myriophyllum alterniflorum ②

Myriophyllum spicatum

Phalaris arundinacea ④

Polygonum amphibium ⑤

Ranunculus fluitans (shorter-leaved species infrequent) ①

Sparganium erectum ③

Small grasses

Mosses

Blanket weed

Variants:
(*a*) *Myriophyllum alterniflorum* and *Myriophyllum spicatum* rarely occur together: *M. alterniflorum* is found in upper reaches, *M. spicatum* in lower reaches
(*b*) If there are two or more *Ranunculus* spp., *R. fluitans* occurs in the more downstream reaches and, if both are found at the same site, *R. fluitans* grows in the deeper water
(*c*) In slower reaches, *Polygonum amphibium* is characteristic
(*d*) In silty (but not too disturbed) reaches, *Sparganium emersum* is characteristic
(*e*) In parts of central Scotland, *Potamogeton natans* hybrids, e.g. *Potamogeton* × *sparganifolius* are important

RESISTANT ROCK STREAMS: E VERY MOUNTAINOUS

Rock: 7
Landscape: E
Streams: i + ii, iii, v
Colour bands: *orange—yellow* to *yellow* (with *green*)

Very mountainous landscapes occur particularly in Snowdonia, the Lake District and parts of the Scottish Highlands, including much of the far north, where the very high rainfall increases water force in the same way as does the steeper topography further south. Land use is predominantly mountain to moorland vegetation, with rough pasture below. Within the regions, mountain streams may locally occur where water force is less (but the substrate is mineral), bog streams on local blanket bogs, and bog rivers (landscape G) where there is both much bog in the catchment and flat-enough topography to allow the deposition of enough peat in the channel. The water is sometimes peat-stained. The very swift normal flow, extremely fierce spate flows, and the consequent bouldery or coarse and unstable substrate usually prevent even mosses from growing well. If macrophytes are present at all they are sparse and few, and most sites have none.

i + ii. Small streams

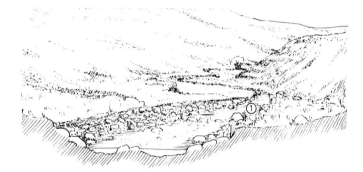

(*yellow*)
0(+) species
0(+)% cover

Local to frequent. 1—4 m wide, *c.* 20—40 cm deep. Perennial flow.

Probable species: ┼ (*Juncus articulatus*) ①   ✹ Mosses

Variant:
(*a*) Dry channels with loose scree and no aquatic species

iii. Medium streams

(*orange—yellow*)
0(+) species
0(+)% cover

Local to frequent. 4—10 m wide, 30—75+ cm deep. More nutrient-poor than (i + ii) as acid peat can, in more downstream and peaty conditions, occur at the sides. Most channels are empty, as in (i + ii), and most of those with plants have solely mosses.

Probable species: 𝕎 (*Callitriche* spp.)   ⊥ (*Glyceria fluitans* with long floating leaves)

Hard sandstone streams

⌐⊨ (*Potamogeton*   Ψ (Small grasses)
      polygonifolius)   —
                                    ᴥ Mosses

(The small grasses (mainly *Agrostis stolonifera*) are found both reaching into the water from the banks and as temporary patches in the centre of the channel washed down from upstream.)

v. Large streams

(*yellow*)
0(+) species
0(+)% cover

Local to frequent. 10–30+ m wide, depth from 0.5 m to deep. Most channels are empty, but the proportion containing vegetation is higher than in (iii).

Probable species:
     Ѱ *Callitriche* spp.   ⅄ *Ranunculus*
                                      (probably *R. aquatilis*)

     Ӿ *Carex acuta*   �室 *Sparganium emersum*

     ⊥ *Juncus articulatus*   Ψ Small grasses
                                    —
     ⋇ *Myriophyllum*   ᴥ Mosses
       alterniflorum

     ⊪ *Phalaris arundinacea*

(Small grasses are found as in (iii) above.)

Variants:
(*a*) With lower water force there can be *c*. 6 species present, the vegetation approaching that of mountain streams
(*b*) With more bog peat, more *orange* species are present
(*c*) With more inorganic silt, more *purple* or *red* species are present

HARD SANDSTONE STREAMS

Rock: 8
Landscapes: C, D, E, (B, G, F)
Streams: i, ii, iii, v
Colour bands: *yellow* to *purple*

Hard sandstone country is mostly upland, though there are some mountain areas and, in central Scotland, lowland and very mountainous topography. In Caithness, in the far north of Scotland, there is lowland country with a peaty mountainous character to the vegetation. In the steeper parts flow is as unstable and fierce as on Resistant rocks, and as sandstone is more erodable, and also produces more silt, the upper streams with the fiercest flows are emptier than those on comparable Resistant rock topography. In the uplands, though, where silt can accumulate, a species-rich, high-cover vegetation often occurs. Compared with lowland soft sandstone streams, however, the silt is washed downstream more, which means *purple* and *red* species occur more in lower reaches, and less in upper ones, than they do in soft sandstone streams. Sand or (in steeper streams) exposed sandstone are common, and both decrease bed stability. The alternating stretches of shallow swift and deep slow water so characteristic of Resistant rock streams are less common on hard sandstone. The variation in possible flow regimes means that, except in clearly defined upland regions and in Caithness, it can be difficult to identify stream type, so the description fitting the site most closely, from the several possibilities, should be used. Damage can be diagnosed only in certain circumstances: see Chapter 4. The nutrient status (nutrient concentrations and hardness ratio) is close to that of chalk. Also see general notes on topography in the section on Resistant rocks (p. 43).

Fringing herbs in upland brooks are like those of chalk in that they are abundant, sometimes in carpets, often in bands and long clumps, and are short. However, because the flows are more eroding than on chalk, submerged carpets are absent. Mosses are less frequent than on any other hard rock (because of the unstable substrates).

Hard sandstone occurs mainly in the Welsh mountains (east), west and north-west England, central and eastern Scotland, and Caithness.

HARD SANDSTONE STREAMS: C UPLAND (AND B LOWLAND)

Rock: 8
Landscapes: C, (B)
Streams: i, ii, iii, v
Colour bands: *blue* to *purple—red*

The land use is typically grassland, arable or orchard, etc. The combination of high silting and fairly low water force means that vegetation can be dense throughout the streams. As always in the hills, however, local variations in topography can lead to areas of more or of less scour, and so to more or less vegetation.

i. Small streams without water-supported species

blue
1+ species
10+% cover

Infrequent. 0.5—3 m wide, summer-dry or with perennial shallow (*c.* 10—30 cm) flow. A little silting. Low banks. Similar to chalk brooks of equivalent size, but with fewer species.

Probable species: *Apium nodiflorum* ①   *Phalaris arundinacea*

Myosotis scorpioides   *Rorippa nasturtium-aquaticum* agg. ②

Hard sandstone streams

🌱 *Veronica beccabunga*   🌿 Small grasses

Variants:

(*a*) Dry shallow grassed channels

(*b*) Sunk ditches, shaded or with tall species such as *Epilobium hirsutum*, *Sparganium erectum*

(*c*) In less fertile ground, dry channels with *Juncus effusus* and perhaps grasses, often with 100% cover

ii. Small streams with water-supported species

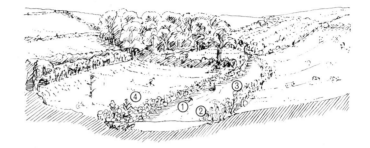

blue (with *green*)
4+ species
40+% cover

Frequent. 2—4 m wide, usually less than 30 cm deep. Slow or moderate flow. Silting at sides and in plant clumps. Banks usually low. Water sometimes peat-stained. Similar to chalk brooks of equivalent size, but without submerged carpets of fringing herbs, and normally without *Ranunculus*.

Often abundant: 🌱 *Callitriche* spp. ①

Often associated: 🌿 *Apium nodiflorum*   🌾 *Sparganium erectum*

🌱 *Mimulus guttatus*   🌱 *Veronica anagallis-aquatica* agg.

🌿 *Myosotis scorpioides* ②   🌱 *Veronica beccabunga* ③

🌾 *Phalaris arundinacea*   🌿 Small grasses

🌱 *Rorippa nasturtium-aquaticum* agg. ④   — 🌿 Mosses

— 🌿 Blanket weed

Variants:

(*a*) In nearly lowland areas, *Elodea canadensis*, *Potamogeton natans*, perhaps *Sparganium emersum* and a selection from the above list are found

(*b*) In swifter streams there is less diversity and lower cover

(*c*) High steep banks with tall plants lead to sparse stream vegetation

iii. Medium streams

blue, *blue—purple* (with *green*)
4+ species
40+% cover

Frequent. 4—9 m wide, usually 30—75 cm deep. Usually with moderate flow, a mixed-grain substrate, and some silting. Water sometimes peat-stained. Not unlike equivalent chalk streams, but recognisably different. Fringing herbs can be fairly large.

Often abundant: 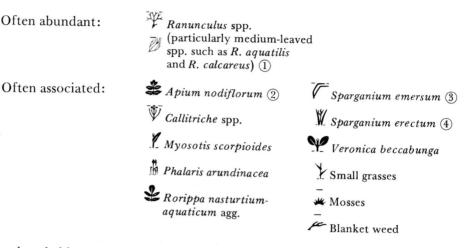 *Ranunculus* spp.
(particularly medium-leaved spp. such as *R. aquatilis* and *R. calcareus*) ①

Often associated:

Apium nodiflorum ② *Sparganium emersum* ③

Callitriche spp. *Sparganium erectum* ④

Myosotis scorpioides *Veronica beccabunga*

Phalaris arundinacea Small grasses

Rorippa nasturtium-aquaticum agg. Mosses

Blanket weed

and probably one or two *purple* or *red* species

Variants:

(*a*) Nearly lowland streams have less *Ranunculus*, and more *Callitriche* spp., *Elodea canadensis*, *Glyceria maxima*, *Potamogeton natans* and *Sparganium emersum*

(*b*) Lower or slower (and therefore siltier) reaches have more *purple* and *red* species, especially:

Elodea canadensis *Sparganium emersum*

Myriophyllum spicatum *Zannichellia palustris*

Potamogeton crispus

(*c*) Swifter streams have lower diversity and cover, and fewer *purple* and *red* species

v. Large streams

blue—purple, purple
(with *green*)
5+ species
25+% cover

Infrequent. 10+ m wide, *c.* 0.5—1+ m deep. Silting occurs. Usually without vegetation in the centre.

Hard sandstone streams

Probable species: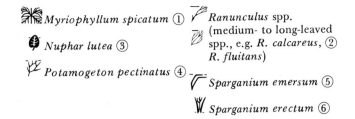
    Myriophyllum spicatum ① Ranunculus spp.
               (medium- to long-leaved
    Nuphar lutea ③   spp., e.g. *R. calcareus*, ②
               R. fluitans)
    Potamogeton pectinatus ④ Sparganium emersum ⑤
               Sparganium erectum ⑥

There is a wide variety of other possible species. The *Ranunculus* species are sometimes dominant.

Variants:
(*a*) In swifter flows there are 4+ species, 10+% to 100% cover, and *Ranunculus* is usually *R. fluitans*
(*b*) In slower flows, mainly in long lowland stretches near the river mouth, there are more *purple* and *red* species:

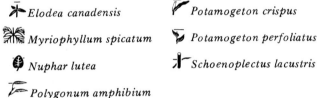

    Elodea canadensis   Potamogeton crispus

    Myriophyllum spicatum Potamogeton perfoliatus

    Nuphar lutea   Schoenoplectus lacustris

and also   Polygonum amphibium

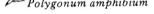

HARD SANDSTONE STREAMS: D MOUNTAIN

Rock: 8
Landscape: D
Streams: i, ii, iii, v
Colour band: *yellow* to *blue—purple*

The land use is, typically, poor grazing. Spatey swift flows, erodable bedrock and incoming silt lead to unstable habitats (particularly in upper reaches) and sparse vegetation throughout.
 Such streams are few, and mainly in South Wales and central Scotland.

i. Small streams without water-supported species

(*green* to *blue*)
0(+) species
0(+)% cover

Local. 1—4 m wide, *c.* 10—30 cm deep. Most channels empty.

Probable species: Myosotis scorpioides Veronica beccabunga

       Petasites hybridus   Small grasses

(The grasses (mainly *Agrostis stolonifera*) are found both reaching into the water from the banks, and as temporary patches in the centre of the channel washed down from upstream.)

ii. Small streams with water-supported species

(*yellow* to *blue*)
0(+) species
0(+)% cover

Local. 1–4 m wide, *c.* 15–40 cm deep. Most channels empty.

Probable species:

Mimulus guttatus   Small grasses

Petasites hybridus   Mosses

Veronica beccabunga

(The small grasses are as in (i) above.)

Variant:
(*a*) *Myriophyllum alterniflorum* in less fertile landscapes

iii. Medium streams

(*yellow* to) *blue*
0+, 2+ species
0+% cover

Infrequent. 4–8 m wide, usually 30–100 cm deep. Channels often empty.

Probable species:

Mimulus guttatus   Small grasses

Sparganium erectum   Mosses

Veronica beccabunga

Rarer water-supported species include:

Sparganium emersum   Blanket weed

Enteromorpha sp.

v. Large streams

blue, blue–purple, with *green*
0+, 5+ species
0+%, 20+% cover

Infrequent. 8–20+ m wide, often over 75 cm deep.

Probable species:

Elodea canadensis   *Sparganium erectum*

Phalaris arundinacea   Small grasses

Ranunculus spp.   Mosses
(mainly *R. fluitans*)

Hard sandstone streams

~ *Enteromorpha* sp.　　　~ Blanket weed

HARD SANDSTONE STREAMS: E VERY MOUNTAINOUS

Rock: 8
Landscape: E
Streams: i + ii, iii
Colour bands:
yellow to *blue*

The very fierce flow and unstable substrate keep most channels empty.
　　Local, mainly in east-central Scotland and South Wales.

i + ii. Small streams

(*yellow*, *blue*)
0(+) species
0(+)% cover

Local. 1—4 m wide, with shallow very eroding flows.

Least-rare species:　　　~ *Veronica beccabunga*　　　~ Mosses

　　　　　　　　　　　　~ Small grasses

iii. Medium streams

(*yellow*, *blue*)
0(+) species
0(+)% cover

Local. 4—10 m wide, 30—75+ cm deep, with eroding flows.

Least-rare species:　　　~ *Callitriche* spp.　　　~ Small grasses

　　　　　　　　　　　　~ *Phalaris arundinacea*　　~ Mosses

　　　　　　　　　　　　~ *Ranunculus* spp.
　　　　　　　　　　　　(short-leaved)

HARD SANDSTONE STREAMS: CAITHNESS (B, F, C)

Rock: 8
Landscapes: B, F, C
Streams: i, ii, iii, (v)
Colour bands: *yellow* to *purple*

i. Small streams without water-supported species

yellow, blue
1+ species
1+% cover

0.5—2 m wide, usually over 30 cm deep. Slow flow.

Probable species:

Carex spp.	*Sparganium erectum*
Juncus articulatus	Small grasses
Phalaris arundinacea	

ii. Small streams with water-supported species

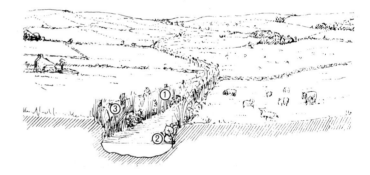

yellow to *blue*
3+ species
20+% cover

1—4 m wide, usually at least 40 cm deep. Slow to moderate flow.

More probable species:

Small grasses	Mosses

Less-probable species:

Callitriche spp.	*Juncus effusus* ①
Caltha palustris ②	*Mentha aquatica*
Iris pseudacorus	*Mimulus guttatus*
Juncus articulatus	*Sparganium erectum* ③

iii. Medium streams

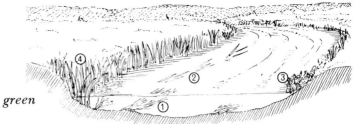

blue, blue—purple with *green*
4+ species
20+% cover

4—8 m wide, usually at least 50 cm deep. Slow to moderate flow.

Probable species:

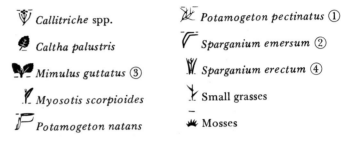

Callitriche spp.

Caltha palustris

Mimulus guttatus ③

Myosotis scorpioides

Potamogeton natans

Potamogeton pectinatus ①

Sparganium emersum ②

Sparganium erectum ④

Small grasses

Mosses

v. Large streams

blue—purple
7+ species
20+% cover

Rare. 10—20 m wide, deep. Varying flow type. Species generally similar to those of (iii).

HARD LIMESTONE STREAMS

Rock: 9
Landscapes: (B), C, D, (E)
Streams: i, ii, iii, v
Colour bands: (*orange*) *yellow* to *blue—purple* (with *green*)

Most hard limestone streams are mountainous. Even though the hills are often well below 2000′ (610 m), the fall from hilltop to stream channel is over 600′ (185 m), giving high force to the water. Silting, as on soft limestone, is low, and so the steeper channels contain negligible silt. There is an unusually great difference between the mountain vegetation, which is very sparse indeed, and the upland vegetation, which is abundant and often similar to that of chalk. Unless a stream is clearly upland, therefore, absence of plants (in a site whose former vegetation is unknown), cannot be assumed to be due to damage. In some parts acid peat can cover hard limestone, and this peat can be deposited in small slow channels, bringing in *orange* or even *brown* species. If the streams are influenced solely by the peat, their vegetation will be that of Resistant rock streams on landscape types F and G. With increasing limestone influence vegetation will be transitional between this and that of pure limestone. In nutrient status, the (non-peaty) silt is close to that of soft sandstone (with higher sodium) and is richer than that of chalk. However, where swift flow leads to coarse substrates, stream nutrient status can be low.

69

Also see the general notes on topography in the section on Resistant rocks (p. 43).

Fringing herbs are usually emerged, not very large, and can be abundant and diverse in small brooks. Mosses are nearly as frequent as on Resistant rocks (and much more so than on hard sandstone). On moorland, etc., dry (or shallow) small channels with *Juncus effusus* are very characteristic.

Hard limestone streams are most common in the Pennines, with smaller areas in e.g. the Mendips, and south and north Scotland.

HARD LIMESTONE STREAMS: C UPLAND (AND B LOWLAND)

Rock: 9
Landscapes: (B), C
Streams: i, ii, iii, (v)
Colour bands: (*orange*) *yellow*
to *blue—purple* (with *green*)

The land use is typically good grassland in the valleys, perhaps with moorland on the hills. As always in the hills, local variations in topography can lead to streams or reaches with greater scour and thus less plants. Substrates are mixed-grained, with much gravel.

i. Small streams without water-supported species

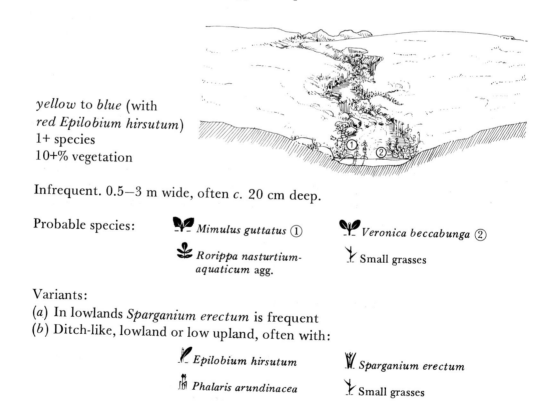

yellow to *blue* (with
red Epilobium hirsutum)
1+ species
10+% vegetation

Infrequent. 0.5—3 m wide, often *c.* 20 cm deep.

Probable species: *Mimulus guttatus* ①   *Veronica beccabunga* ②

Rorippa nasturtium-aquaticum agg.   Small grasses

Variants:
(*a*) In lowlands *Sparganium erectum* is frequent
(*b*) Ditch-like, lowland or low upland, often with:

Epilobium hirsutum   *Sparganium erectum*

Phalaris arundinacea   Small grasses

(*c*) Peaty, also with perhaps:

Caltha palustris

Eleocharis palustris

Juncus articulatus

Iris pseudacorus

(*d*) Peaty dry channels, with *Juncus effusus* dominant or with other species
(*e*) Empty channels, often dry but with scouring storm flows, and with stony or peaty substrates
(*f*) Small dry shallow grassy channels

ii. Small streams with water-supported species

(*orange* to) *blue*
3+ species
40+% cover

Frequent. 1—4 m wide, *c*. 20—50 cm deep. Some sediment in plant clumps, etc.

Probable species:

Mimulus guttatus ①

Myosotis scorpioides

Rorippa nasturtium-aquaticum agg. ②

(Sparganium erectum)

Veronica beccabunga

Small grasses

Mosses

Blanket weed ③

Variants:
(*a*) Peaty, also with perhaps:

Caltha palustris

Juncus articulatus ③

Phalaris arundinacea

Ranunculus flammula

(*b*) More water force and/or a mixture of peaty and limestone substrate produces little vegetation, with mosses the commonest species
(*c*) In a low-rainfall area without spate there may be 5+ species and 80+% cover. Often abundant are:

 *Ranunculus* spp. (mainly *R. aquatilis*, *R. calcareus*)

and often associated are:

Apium nodiflorum

Callitriche spp.

Rorippa nasturtium-aquaticum agg.

Mosses

Blanket weed

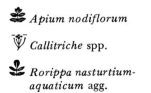

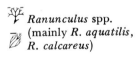

iii. Medium streams

blue (to *purple*, with *green*)
4+ species
50+% cover, but 20+%
with more spate

Frequent. 4—8 m wide, *c*. 30—75 cm deep. Usually moderate flow.

Often abundant: 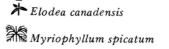 *Ranunculus* spp. ①
(short- to medium-leaved)

Often associated:

(*Groenlandia densa*)	*Veronica beccabunga*
Mentha aquatica ②	Small grasses
Myosotis scorpioides ③	Mosses
Rorippa nasturtium- *aquaticum* agg. ④	Blanket weed

Variants:

(*a*) Downstream, deeper reaches also have *purple* (and *red*) species: 6+ species in all. These may include:

Elodea canadensis	*Zannichellia palustris*
Myriophyllum spicatum	*Enteromorpha* sp.

(*b*) When there is slow flow, but high rainfall and some sedimentation, there will only be 2+ species in all

Often abundant:

Potamogeton natans	*Sparganium erectum*
Sparganium emersum	

Often associated:

Callitriche spp.	Mosses
Small grasses	

(*c*) Peaty streams without high water force have 2+ species in all, and also perhaps:

Carex spp.	*Phalaris arundinacea*
Juncus articulatus	

(*d*) With more water force and/or a mixture of peaty and limestone substrate there is little vegetation, mosses are the most frequent species, and small grasses and blanket weed not rare

v. Large streams

blue to *blue–purple*
5+ species
40+% cover

Rare. 8–20+ m wide, 0.5 m and deeper. Typically with moderate flow. Species list uncertain but probably similar to (iii).

GENERAL VARIANT: MAGNESIAN LIMESTONE

Rock: 9
Landscapes: B, (C)
Streams: i, (ii), iii
Colour bands: *blue* to ?

Land use arable and grassland. Undamaged vegetation uncertain as the only large river on the rock is seriously polluted, and the other streams tend to be either small and shaded, etc., or large but on the rock for too short a distance to be affected by it.

Magnesian limestone outcrops in a narrow band along the eastern edge of the Pennines, and in a patch north of Teesside, in lowland or nearly-lowland country.

i. Small streams without water-supported species

blue
3+ species?
30+% cover

1–3 m wide, shallow (e.g. 20 cm) to summer-dry. Slow flow, usually silty. Vegetation often damaged.

Probable lowland species: *Apium nodiflorum*   *Solanum dulcamara*

Epilobium hirsutum   *Sparganium erectum*

Glyceria maxima

Upland species are not known.

(ii) + iii. Larger streams

Their undamaged vegetation is not known.

HARD LIMESTONE STREAMS: D MOUNTAIN (AND E VERY MOUNTAINOUS)

Rock: 9
Landscapes: D, (E)
Streams: (i), ii, iii, v
Colour bands: *yellow*
to *blue*, with *green*

Land use is often good grassland in the valleys, with poor grassland and mountain or moorland vegetation above. The fall from hilltop to stream channel must be carefully checked, as mountain limestone streams can occur with hills only *c.* 1000′ (305 m) high, even if these are found only at the head of the catchment. Local variations in topography can lead to upland tributaries in flatter areas (or very mountainous ones in steeper areas, etc.). Intermediates occur between mountain and upland streams. Exposed rock (or boulders) is common on the bed, where the swift flow and fierce spates combine with the unsuitable substrate to make the habitat unfavourable for vegetation. Alternating stretches of swifter shallower (often rapid) water and slower deeper water are common.

(i) + ii. Small streams

(*yellow* to *blue*, with *green*)
0+ species
0+% cover

Frequent. 0.5—4 m wide, *c.* 10—30 cm deep. Most channels empty. Channels with plants usually have mosses, whether with or without emergents.

Probable species: (*Petasites hybridus*)   Mosses

     (*Veronica beccabunga*)   (Blanket weed)

     (Small grasses)

Variants:
(*a*) Peaty, perhaps with *Juncus articulatus* or *Eleocharis palustris*
(*b*) Peaty dry channels, with *Juncus effusus* dominant or with other species
(*c*) Small dry shallow grassy channels (and grading to landscape type B, upland)

74

Hard limestone streams

iii. Medium streams

(*yellow* to *blue–purple*,
with *green*)
0+ species
0+% cover

Frequent. 4–8 m wide, usually 20–75 cm deep. Many channels empty.

Probable species: 🌿 (*Petasites hybridus*) ①   🌱 Mosses

🌾 (Small grasses)   〰 (Blanket weed)

Variant:

(*a*) In lower reaches, or other stretches with lower water force, the following are
infrequently found:

🌾 *Carex acutiformis*   🌿 *Rorippa nasturtium-
aquaticum* agg.

✳ *Elodea canadensis* ③

🌱 *Ranunculus* spp.
(short-leaved)

🌾 *Phalaris arundinacea* ②

🌾 *Sparganium erectum*

🌿 *Zannichellia palustris*

v. Large streams

yellow (to *blue–purple*, with *green*)
(0+), 1+ species
0+% cover

Infrequent. 10–20+ m wide, *c.* 30 cm– 2 m deep. Some channels empty.

Probable species: ✳ (*Elodea canadensis*)   🌾 Small grasses

🌾 (*Phalaris arundinacea*)   🌱 Mosses

🌱 (*Ranunculus* spp.,
short to medium-leaved)   〰 Blanket weed

Variant:

(*a*) Slower parts may have more vegetation, and perhaps also:

🌾 *Eleocharis palustris*   🌿 *Potamogeton pectinatus*

🌿 *Mimulus guttatus*   🌾 *Sparganium erectum*

🌿 *Myriophyllum spicatum*   🌿 *Veronica beccabunga*

75

CALCAREOUS AND FELL SANDSTONE STREAMS

Rock: 10
Landscapes: C, (B)
Streams: i, ii, iii, (v)
Colour bands: *blue* to *purple*, with *green* (with *yellow* and *orange* on peaty catchments)

The vegetation of these streams is intermediate between those on hard sandstone and those on hard limestone. The landscape is usually upland, but lowland in parts. Because of the combination of lower water force (upland—lowland, not mountain), high silting (sandstone) and high rock fertility (lime), the vegetation is more slanted towards *purple* than in the sandstone or limestone streams. Silt and sand are common on the bed.

In some parts, however, particularly uplands, peaty moor overlies the landscape. This greatly reduces the fertility of the streams, and the vegetation, and slants the plant community towards *yellow*.

The North Yorks Moors have both sandstone and limestone outcrops, and the streams are mountain streams, but the less calcareous ones can best be classified here, as species-poor members of the peaty variants below.

Calcareous sandstone streams occur in north-east England and south-east Scotland.

i + ii. Small streams

blue
4+ species
25+% cover

Frequent. 1—4 m wide, usually 20—50 cm deep.

Probable species:

    Callitriche spp.

    Elodea canadensis

    Myosotis scorpioides

    Rorippa nasturtium-aquaticum agg.

    Sparganium erectum

    Veronica anagallis aquatica agg.

    Veronica beccabunga

    Mosses

    Blanket weed

Variant:
(*a*) Dry shallow channels

Peaty variants:
(*a*) Dry or shallow-water channels with abundant *Juncus effusus* and perhaps other species
(*b*) Empty channels, often dry, stony or with swift flow
(*c*) Locally frequent channels, 20—30 cm deep, with fast flow, peat-stained water and stony substrate.

Calcareous and fell sandstone streams

Probable species: 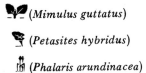 (*Mimulus guttatus*)   (Small grasses)

 (*Petasites hybridus*)   Mosses

 (*Phalaris arundinacea*)   (Blanket weed)

Other *yellow* or *orange* species may be present

iii + (v). Medium (to large) streams

blue—purple, with *green*
6+ species
25+% cover

Frequent. 5—15 m wide, usually 30 cm to 1 m or more deep.

Probable species:

 Elodea canadensis   *Sparganium emersum*

 Myosotis scorpioides   *Sparganium erectum*

 Phalaris arundinacea   *Veronica beccabunga*

 Polygonum amphibium   Mosses

 Potamogeton natans   *Enteromorpha* sp.

 Ranunculus spp.   Blanket weed

 Schoenoplectus lacustris

Peaty variants:
(*a*) Locally frequent, usually 25—60+ cm deep, with fairly swift flow and peat-stained water, substrate stony or mixed-grained. Diversity variable but usually low; cover low. Probable species:

 Carex acutiformis   *Ranunculus* spp.

 Eleocharis palustris   *Sparganium erectum*

 (*Epilobium hirsutum*)   Small grasses

 Myosotis scorpioides   Mosses

 Phalaris arundinacea

(*b*) Local, with less peat than above (sometimes downstream of the above) and intermediate in habitat and vegetation. Colour bands are (*yellow—*) *blue*, there are 5+ species and 40+% cover. Probable species:

 Epilobium hirsutum   *Sparganium erectum*

 Mimulus guttatus   Small grasses

 Myosotis scorpioides   *Enteromorpha* sp.

 Petasites hybridus   Blanket weed

 Phalaris arundinacea

COAL MEASURES STREAMS

Rock: 11
Landscapes: C, D
Streams: i, ii, iii, v
Colour bands: *yellow*
to *purple*, with *green*

The landscape is upland or mountain (where Coal Measures outcrop in lowland areas, as in parts of the English Midlands, they are classed as clay). The substrate is often fairly coarse, because of the high water force in steeper areas. The nutrient status of the silt is high, fairly close to that of clay, so the streams are nutrient-rich where low water force permits silt to accumulate. Peaty moorland may overlie the rock in some parts, and influence the vegetation.

Coal Measures streams are usually polluted from mine and industrial effluents, so the undamaged vegetation is difficult to ascertain, and that described here may be incomplete.

Also see the general notes on topography in the section on Resistant rocks (p. 43).

If damage ratings are wanted in slower-flowing streams, use the higher values of diversity and cover when two are given (see Chapter 4).

Coal Measures streams are most common in South Wales and the southern Pennines, but also occur elsewhere, e.g. southern Scotland.

i + ii. Upland and lowland small streams

blue to *purple* (with *green*)
2+ species
20+% cover

Infrequent. 1—4 m wide, *c.* 10—40 cm deep. Usually have slow flow and little spate.

Probable species:

Alisma plantago-aquatica

Callitriche spp.

Elodea canadensis

Lemna minor agg.

Potamogeton crispus

Potamogeton natans

Sparganium emersum

Sparganium erectum

Veronica beccabunga

Small grasses

Blanket weed

Variants:
(*a*) Swifter flow, less vegetation
(*b*) Lowland streams with less water force have 4+ species, 40+% cover
(*c*) Ditch-type variants have tall emergents and may be summer-dry, with e.g.

Epilobium hirsutum   *Sparganium erectum*

(*d*) Dry shallow grassy channels

i + ii. Mountain, and swifter upland small streams

(*yellow*)
0+ species
0+% cover

Infrequent. 1—4 m wide, *c.* 20—40 cm deep. Channels often empty.

Probable species:

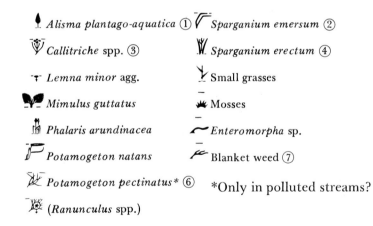

Myosotis scorpioides   Mosses

(Small grasses)   (Blanket weed)

and other species from (i + ii) above.

Variant:
(*a*) More peaty. Probable species, as well as the above, include:

Eleocharis palustris   *Petasites hybridus*

Juncus articulatus   *Phalaris arundinacea*

iii + v. Medium and large streams

yellow to *blue—purple*
3+, 8+ species
10+%, 25+% cover

Frequent. 4—30 m wide, depth from 30 cm to deep. Clean vegetation dubious. Less, and *yellower* vegetation in swifter streams; more, and more *purple* vegetation in slower ones.

Probable species?:

Alisma plantago-aquatica ①   *Sparganium emersum* ②

Callitriche spp. ③   *Sparganium erectum* ④

Lemna minor agg.   Small grasses

Mimulus guttatus   Mosses

Phalaris arundinacea   *Enteromorpha* sp.

Potamogeton natans   Blanket weed ⑦

*Potamogeton pectinatus** ⑥   *Only in polluted streams?

(*Ranunculus* spp.)

STREAMS ON MIXED CATCHMENTS (HARD ROCKS)

Rocks: 7, 8, 9, 10, 11
Landscapes: C, D, E (etc.)
Streams: (i, ii), iii, v
Colour bands: *yellow* to *blue-purple* (with *green*)

A mixed catchment is one which, upstream of the site recorded, consists of two or more rock types. Vegetation usually reflects the proportions of each, subject to the following rules:

1. Rocks near the headwaters are the most important, and those at the side near the mouth the least so.

2. When the proportions of the different rock types are equal, the rock at the site is the more important.

3. Hard sandstone has a stronger influence than hard limestone, which in turn has a stronger influence than Resistant rocks or Coal Measures.

4. When sandstone and limestone are either or both of the rock types concerned, deeper slower flows, allowing silting, lead to markedly more-eutrophic vegetation, and swifter shallower ones to markedly less-eutrophic vegetation.

5. Drift (Boulder Clay, etc.) is important only when thickly covering large proportions of the catchment, as in the Boulder Clay of Anglesey.

6. The change in plant community after crossing a rock boundary occurs much sooner if tributaries enter which rise on the new rock type, or if the new rock type is sandstone.

7. The rock type map (pp. 126–45) is only on a small scale. River plants are very sensitive to rock type, and when vegetation (*a*) does not fit that expected from the rock type map, and (*b*) is species-rich and does not show a characteristic damage pattern (see Chapter 4), a large-scale geological map should be consulted. There will probably be a small patch of, say, sandstone in a Resistant rock area.

8. As streams on hard rocks have lower species diversity and less total vegetation than those on soft rocks, the effects of changes in rock are less, and are less easy to detect. The vegetation of streams on hard rocks also alters greatly with topography. Within Resistant rock areas, for instance, granite and basalt form flatter landscapes, and their streams consequently contain more *Callitriche* species and *Glyceria maxima*.

Necessarily, most streams flowing on two rock types are quite large. The few small ones occur where there is a mosaic of rock types (e.g. hard sandstone and Resistant rock).

HARD SANDSTONE–RESISTANT ROCK STREAMS

Rocks: 7, 8
Landscapes: C, D, (E)
Streams: (i, ii), iii, v
Colour bands: *yellow* to *blue–purple* (with *green*)

Derive the expected vegetation from that of hard sandstone and that of Resistant rock, in accordance with rules 1 to 8 above and:
1. Streams on mixed catchments have more

       Elodea canadensis   *Sparganium emersum*
       Potamogeton natans

than streams solely on one of the rock types.
2. When considering catchments with different proportions of the two rocks:
   (*a*) in streams rising on Resistant rock and flowing on to sandstone, the following decrease with decreasing sandstone:

       Callitriche spp.   *Sparganium erectum*
       Potamogeton crispus

Streams on mixed catchments (hard rocks)

(*b*) In (the fewer) streams which rise on sandstone and flow on to Resistant rock, with decreasing sandstone there is an additional decrease in:

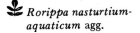 *Rorippa nasturtium-aquaticum* agg.

(*c*) Mosses increase with decreasing proportions of hard sandstone.

HARD LIMESTONE–RESISTANT ROCK STREAMS

Rocks: 7, 9
Landscapes: (C), D
Streams: (i, ii), iii, v
Colour bands: *yellow* to *blue*, with *green*

Derive the expected vegetation from that of hard limestone and that of Resistant rock, in accordance with rules 1 to 8 above and:
1. In catchments with different proportions of the two rocks, mosses decrease with decreasing proportions of Resistant rock.

CALCAREOUS SANDSTONE–RESISTANT ROCK STREAMS

Rocks: 7, 10
Landscape: C
Streams: (iii, v)
Colour bands: *yellow* to *blue–purple* (with *green*)

These streams are rare. Their vegetation is as expected from rules 1 to 8 above, but with
more:

Elodea canadensis

Myosotis scorpioides

Ranunculus spp.

Sparganium erectum

Enteromorpha sp.

less:

Mosses

However, the streams are too few for it to be certain whether this difference is due to the mixed catchment or merely to streams on mixed catchments being, in general, larger than those on calcareous sandstone alone.

COAL MEASURES–RESISTANT ROCK STREAMS

Rocks: 7, 11
Landscapes: C, D
Streams: (i, ii, iii, v)
Colour bands: *yellow* to *blue–purple*, with *green*

Derive the expected vegetation from that of Coal Measures and that of Resistant rock, in accordance with rules 1 to 8 above. However, both types bear little vegetation, and most large Coal Measures streams are polluted.

4 Stream damage rating and pollution index

INTRODUCTION

The damage rating assesses the amount of damage to a site, from any cause — whether shading, herbicides or anything else. It is measured on an eight-point scale from *a* (all right) to *h* (horrible). When the damage can be attributed to a chemical cause, i.e. where there is evidence of chemical damage, a pollution index can be assigned, again on an eight-point scale but using capital letters *A* to *H* and also with *U* for unclassable sites. The pollution index has been slanted towards assessing the damage from town (plus industrial) effluents, which are the most important and widespread form of river pollution in Britain.

Only major damage to the quality of the vegetation is assessed by this damage rating. It is possible to decrease biomass without decreasing diversity, cover, or colour band, etc. — as in light shade, or water which is somewhat turbid but with no chemical influence — and that sort of damage is not measured by this method. There is frequently, in the lowlands, considerably more cover and diversity than is given in the description in Chapter 3, and it is hoped, in future, to subdivide the *a/A* category to take account of this. At present, however, it is not possible, in sites of unknown history, to separate the 'natural' from the 'damage' factors causing these variations.

The basis on which a damage rating is assigned to a site is the difference between the actual vegetation recorded and the expected vegetation as described in Chapter 3. If the two are the same, a damage rating of *a* and a pollution index of *A* can be assigned automatically. If they are different, the differences should be noted and applied in Table 4.1. A list of pollution-tolerant species — which also are used to assess the rating — is given in Table 4.2.

ASSESSMENT OF DAMAGE RATING

Choice of sites

The instructions given in Chapter 1 must be followed carefully. If several sites in a similar habitat differ in vegetation, all can be used for damage ratings, but only that with the most vegetation for the pollution index (for provided the habitat is uniform then the other sites must have been damaged by factors other than pollution).

Applicability of the damage rating

The damage rating cannot be used where very little vegetation is expected, or when the vegetation cannot be predicted accurately. Its applicability thus varies according to rock type and stream size, as described below.

82

Assessment of damage rating

Lowlands: soft rocks

(a) It can be used in stream sizes (ii), (iii) and (iv) on catchments of a single rock type.

(b) It is less accurate in streams of size (i), as these usually have low diversity or cover and so departures from these are less easy to give accurately.

(c) It is also less accurate in stream sizes (ii), (iii) and (iv) on catchments of mixed rock type, as here it is less easy to predict vegetation. Accuracy is possible if either undamaged vegetation can be recorded in a uniform habitat, and taken as the 'expected' vegetation, or the botanist is sufficiently experienced to know, from the rock types and landscape pattern of the particular catchment, what the expected vegetation should be.

Lowlands: hard rocks

It is less accurate on hard rocks than on soft rocks as it is less easy to predict vegetation. This is partly because it varies more in different parts of the country, and partly because small changes in topography can more easily alter silting (nutrient status) and water force. Alternatively more fertile landscapes (e.g. arable) have more eutrophic stream vegetation. Accuracy is possible where, in a uniform habitat, undamaged and damaged vegetation can both be recorded, either in the same stream or in neighbouring streams. The undamaged vegetation is then taken as the 'expected' community.

Uplands

(a) It can be used in stream sizes (ii), (iii) and (v) on catchments of a single rock type where considerable vegetation is expected. The notes for lowland hard rocks (above) also apply here.

(b) It is less accurate in streams where little vegetation is expected, whether this is due to smallness, steepness or rock type.

(c) It is also less accurate in streams on mixed rock types (for the same reason as given in the section on lowland soft rocks above).

(d) Trends can be seen and differences in damage assessed in streams with little vegetation, if marked changes in vegetation occur between surveys in different years.

Mountains (and streams rising in mountains)

(a) It can be used with fair accuracy where undamaged reaches have considerable vegetation, and the damaged reaches have similar water force and nutrient regime to the undamaged reaches — and, preferably, are downstream of them. In this case, the undamaged vegetation should again be taken as the 'expected' community, and the damaged one assessed relative to this.

(b) It gives only poor accuracy in streams where *Potamogeton pectinatus* is the main or the only species. These are seriously polluted; probably *F* (see below) if without other species or *E* if other species are present.

(c) It gives fair to poor accuracy in streams with changing vegetation (see section on uplands, above).

Notes on the use of damage ratings

Numbers refer to sections in Table 4.1.

1. No comments.

2. If there are more species present than are expected, the figure assigned is 0, not a negative number.

4 Stream damage rating and pollution index

Table 4.1. *Stream damage rating*

The expected species diversity, percentage cover and colour band for each stream type are given in Chapter 3, and are used in sections 2, 3 and 4 below. The lowland silt weighting, section 7, is obtainable from the rock type and stream size (see Chapter 2).

1. Species diversity allowance*

No. species present	0	1—2	3—4	5—6	7—8	9+
Assign figure of:	5	4	3	2	1	0

2. Decrease in diversity: difference between expected and actual number of species*

3. Percentage decrease in percentage cover

Percentage loss in cover

in water up to 1 m deep:	100	80—95	60—75	40—55	20—35	0—15
Assign figure of:	5	4	3	2	1	0

4. Change in colour band

Change:	More than one band	One band or change to uncertain or nil	Half band	Dubious change	No change
Assign figure of:	4	3	2	1	0

5. Percentage of pollution-tolerant species

See Table 4.2. Add 1 for each tolerant species (and for land species rooted in streams iii—v), ½ for each semi-tolerant one

	Nil species	100% tolerant	75—95 %	50—70 %	30—45 %	15—25 %	0—10 %
Assign figure of:	5	5	4	3	2	1	0

Add 4 if only sensitive species are present but the number present is not over one-sixth of those expected and 3 if one quarter (or approx.)

6. Weighting for *Potamogeton pectinatus* and blanket weed

	Much *Potamogeton pectinatus*	Sparse *Potamogeton pectinatus* the only species present	Much blanket weed
Assign figure of:	4 (2 if intermediate)	1	2

7. Weighting for organic silt (lowlands only)

Clay (iv)
Clay-mix with much clay (iv)   Much organic silt: subtract 2

Flatter sandstone (iii)—(iv)
Clay (iii), slower   Less organic silt: subtract 1
Clay-mix (iv), or, if flat, (iii)

Add 4 if only sensitive species are present but the number present is not over one-sixth of those expected

Total	Damage rating
0—4	a
5—7	b
8—10	c
11—13	d
14—16	e
17—18	f
19—21	g
22+	h

If channels with over 15% cover are assessed as *h*, they are automatically reclassified as *g*

*Mosses restricted to man-made structures (bridge piers, concrete slopes etc.) should be disregarded.
'Species' as defined on p. 16.

Assessment of damage rating

Table 4.2. *Pollution-tolerant species*

TOLERANT:	*Potamogeton crispus*	*Sparganium erectum*
	Potamogeton pectinatus	*Enteromorpha* sp.
	Schoenoplectus lacustris	Blanket weed
	Sparganium emersum	
In lowland streams (i) and (ii):	*Apium nodiflorum*	
On hilly Resistant rock, Coal Measures and hard limestone:	*Mimulus guttatus*	
SEMI-TOLERANT:	Small grasses*	*Lemna minor* agg.
	Butomus umbellatus	*Nuphar lutea*
	Glyceria maxima	*Rorippa amphibia*
Limestone and Resistant rock, Coal Measures and hard limestone:	*Phalaris arundinacea*	
On hilly Resistant rock:	*Ranunculus* spp.	

*Excluding *Glyceria fluitans* with long floating leaves (and *Catabrosa aquatica*).

If there are substantially more species present in the hills than are listed as expected, and there is no topographical reason for low water force, then the water force has probably been lessened by a reservoir, weir, or other structure (see Chapter 6).

3. Note that it is the *percentage decrease* in percentage cover, not the *change* in percentage cover, which is measured. The cover is that in water up to 1 m deep, or, in deep or turbid channels, the cover at the sides. If there is more cover than expected the figure assigned is 0, not a negative number. Land species, if present, are ignored when assessing cover.

Ignore cover from *Lemna* species of over 10%.

If cover is substantially higher in the hills than that listed as expected, and there is no topographical reason for low water force, then the water force has probably been lessened by a reservoir, weir, or other structure (see Chapter 6).

4. A change of one colour band is, e.g., from *blue* to *purple*. A change of half a colour band is, e.g., from *blue–purple* to *purple*. A change to an uncertain colour band, or to none, is a change from an expected definite colour (see Chapter 2 and the lists in Chapter 3) to an empty channel or an unclear colour (e.g. grass, *Lemna minor* agg. and blanket weed).

This section assesses change in trophic status, whether an increase in nutrients (eutrophication) or a decrease.

When more than one colour band is listed in Chapter 3 for one stream type, the colour moves towards *red* with slow silty downstream conditions and towards *yellow* with the reverse. The correct band can soon be judged with experience, and if beginners assign a figure of, say, 4 instead of 3, then this makes only a slight difference to the final result.

5. Example: *Agrostis stolonifera, Groenlandia densa, Myosotis scorpioides, Potamogeton crispus, Rorippa nasturtium-aquaticum* agg., blanket weed. Six species, of which two are tolerant (score 2) and one semi-tolerant (score ½). Total: 2½ out of 6, or *c.* 40%. Assign figure of 2.

The pollution-tolerant species are those tolerant to town effluents. In this respect the damage rating is more slanted towards this type of damage than to other forms of (non-nutrient) pollution, the reason being that town effluents are (1979) the most widespread type of pollution. However, the rating can be modified in this section — and in 7 — to assess any other form of pollution, provided that the species tolerant and sensitive to it are known.

6. This weighting is necessary because these species may actually increase with certain forms of damage (particularly pollution and salt for *Potamogeton pectinatus*, and dredging, pollution and intensive cutting for blanket weed). Consequently a site may be seriously damaged but have no decrease in plant cover because these species have increased sufficiently to compensate for the decline in sensitive species. An allowance must therefore be made for this.

7. Most species tolerant to pollution are tolerant because they can grow well in organic silt. Therefore, clean sites which bear much organic silt also bear a high proportion of pollution-tolerant species — and must not be diagnosed as polluted. This section adds the appropriate compensating factor.

Empty streams of sizes (i) or (ii) may rate as less than *h* because channels with little vegetation can have this vegetation removed by less severe damage than it takes to remove the greater amount of vegetation found in larger streams. The rating (and, where relevant index) should be given with a minus sign after the letter, to show the damage could be worse, i.e. *f*— means *f, g* or *h*.

Worked examples of damage ratings

Example 1

Four sites downstream of Rugby sewage works (R. Avon), showing recovery from pollution. All on lowland clay.

(a) iii, 4 spp., 80% cover, *purple—red*
    Much: *Potamogeton pectinatus*, blanket weed
    Little: *Callitriche* sp., *Sparganium erectum*
(b) iii—iv, 6 spp., 60% cover, *purple—red*
    Much: *Potamogeton pectinatus*
    Little: *Glyceria maxima, Schoenoplectus lacustris, Sparganium emersum, Sparganium erectum*, blanket weed
(c) iii—iv, 7 spp., 40% cover, *purple—red*
    Much: Blanket weed
    Little: *Iris pseudacorus, Nuphar lutea, Potamogeton pectinatus, Schoenoplectus lacustris, Sparganium emersum, Sparganium erectum*
(d) iv, 11 spp., 40% cover, *purple—red*
    Much: *Nuphar lutea, Sparganium emersum*
    Little: *Callitriche* sp., *Lemna minor* agg., *Phragmites communis, Sagittaria sagittifolia, Schoenoplectus lacustris, Sparganium erectum, Typha latifolia, Enteromorpha* sp., blanket weed

Assessment of damage rating

	Site			
	(a)	(b)	(c)	(d)
1. Diversity allowance	3	2	1	0
2. Diversity decrease	3	2	1	0
3. Cover decrease	0	0	1	1
4. Colour change	1	1	1	0
5. Tolerant species	4	4	4	3
6. Special species	6	4	2	0
7. Silting	−1	−1	−1	−2
Total	16	12	9	2
Rating	*e*	*d*	*c*	*a*

Site (d) has in fact the vegetation appropriate to a clean wide river well downstream, while it is sited at the very upstream end of a size (iv) stream. This leads to an anomaly of a type inevitable in a simple rating scheme: the site is not fully recovered from damage, but only a botanist used to distinguishing between the upper and lower reaches of size (iv) streams can prove this. The damage is slight, and necessarily falls within that allowed for in an undamaged stream (see Preface and above).

Example 2

Site near source of R. Witham, showing, over several years, recovery from pollution and damage from trampling. Lowland oolite, size (ii).
   (a) No spp., badly polluted
   (b) 3 spp., 20% cover, *blue–purple*
       Little: *Rorippa nasturtium-aquaticum* agg., *Veronica beccabunga*, blanket
           weed
   (c) 2 spp., 20% cover, *blue*
       Little: *Rorippa nasturtium-aquaticum* agg., *Veronica beccabunga*
   (d) 6 spp., 50% cover, *blue (purple)*
       Little: *Lemna minor* agg., *Myosotis scorpioides*, *Rorippa nasturtium-
           aquaticum* agg., *Veronica beccabunga*, *Enteromorpha* sp., blanket
           weed
   (e) 2 spp., 10% cover, *blue*, trampled
       Little: *Phalaris arundinacea, Veronica beccabunga*

	Year				
	(a)	(b)	(c)	(d)	(e)
1. Diversity allowance	5	3	4	2	4
2. Diversity decrease	4	1	2	0	2
3. Cover decrease	5	3	3	0	4
4. Colour change	3	2	0	1	0*
5. Tolerant species	5	2	0	2	0
6. Special species	0	0	0	0	0
7. Silting	0	0	0	0	0
Total	22	11	9	5	10
Rating	*h*	*d*	*c*	*b*	*c*

*Note that *Phalaris*, occurring equally in two places on the stream dial, does not alter a colour band lying between those two places.

In only two of these five years were water-supported species present, confirming the correct placement of this site as size (ii). If the site had been recorded in years (a), (c) and (e) only, the initial reaction would have been to place it as size (i). Reading the description, however, it appears that these size (i) streams are dry for most of the summer, so this site, with 20—30 cm water, is properly placed in size (ii) (see Chapter 2 for the placing of streams by depth rather than by width when the two do not correspond).

If the site is classed as size (i), however, the ratings are as follows:

	Year		
	(a)	(c)	(e)
1. Diversity allowance	5	4	4
2. Diversity decrease	3	1	1
3. Cover decrease	5	2	3
4. Colour change	3	0	0
5. Tolerant species	5	0	0
6. Special species	0	0	0
7. Silting	0	0	0
Total	21	7	8
Rating	*g*—*	*b*	*c*

*g— means the site is at least as bad as *g*, i.e. is *g* or *h*.

The ratings for years (a) and (e) are the same in both cases (since *g*— = *g* or *h*); that for year (c) is one class out.

Habit changes with pollution

If any of the following species are yellow or unusually flaccid, pollution can be suspected, even if the site otherwise appears clean. If the damage rating is *a*, the pollution index (for details, see below) should be provisionally given as *B*.

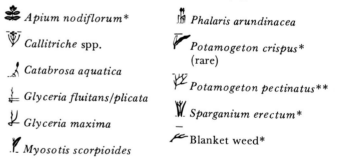

*Apium nodiflorum**	*Phalaris arundinacea*
Callitriche spp.	*Potamogeton crispus** (rare)
Catabrosa aquatica	*Potamogeton pectinatus***
Glyceria fluitans/plicata	
Glyceria maxima	*Sparganium erectum**
Myosotis scorpioides	Blanket weed*

*Also grows large when the stream is polluted
**Also grows very large when the stream is polluted

Odd results

Odd results arise, of course, from many causes, most of them being the standard sources of damage listed in the section on pollution index below. However:
1. On a small-scale rock type, like the one in this book, very small outcrops have to be omitted, though they can be locally important, mainly to small (size i) streams. The presence of an unidentified outcrop can be suspected if: the veg-

etation is as expected in diversity and cover and is not unduly pollution tolerant, but differs in species composition; the type occurs locally on an area otherwise conforming to the vegetation described in Chapter 3 and is confined mainly or entirely to small streams.

The correct diagnosis can be made, if the difference is in rock type, from a large-scale geological map, or if it is in soil type (normally blanket bog) by the fact that the vegetation shifts towards *brown* and the obvious presence of acid peat. Locally thick patches of Drift Clay are the most likely source of rock type variation in lowland Britain.

2. Rainfall is higher in the west and far north than in the east. Thus if streams in hilly western and far northern areas bear consistently less vegetation than expected, particularly if there is a shift in colour towards *yellow* and the change occurs in small streams, rainfall and flood patterns should be checked before damage from causes other than fierce flow is assumed.

3. As a corollary to point 2, if too little vegetation is present the topography, etc., should be checked to see whether the site has the correct vegetation for an area of steeper, higher hills.

4. In somewhat polluted streams (with a pollution index *B—F*) which have intermittent silting, reaches with much silt will be more polluted, with a higher damage rating and pollution index, than those with less silt (because of the pollutants contained in the silt).

5. If the site is, according to the lists in Chapter 3, deficient in emergents (whether fringing herbs or tall monocotyledons), the slope (and type, if concreted etc.) of the banks should be compared with that of the appropriate illustration in Chapter 3. Unduly steep banks lead to loss of emergents.

6. Unduly gentle banks with short or little vegetation may lead to increased emergents and a too-high rating.

ASSESSMENT OF POLLUTION INDEX

Sources of damage other than pollution

The site should be examined for the following:
1. Substantial shade at sides or over whole channel

2. Visitor trampling, paddling or swimming

3. Cattle disturbance, trampling or grazing

4. Boats

5. Recent dredging

6. Recent cutting

7. Herbicides sprayed on emerged (or floating) species

8. Aquatic herbicides used in the water of the channel (1979, in dykes, drains and canals only)

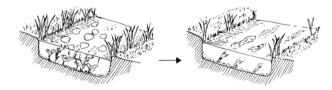

9. Roadworks affecting channel or temporarily causing extra mud, etc., to wash into channel

10. Bed made of concrete, boulders or other coarse substrates

11. Bed of man-made unstable substrate

12. Undue turbulence or deep water caused by bridge piers or other structures

13. Unduly steep banks for the type of channel (see above, and illustrations in Chapter 3)

14. Unduly shallow or wide (if flow swift enough to cause scour), or unduly deep (see illustrations in Chapter 3)

15. In dykes, etc., particularly, substantial lowering of water level during previous year

Severe drought, e.g. as in 1976, can cause damage in streams both in the same and the next year. Also, dyke levels are frequently lowered in winter, and occasionally become sufficiently low to affect the vegetation the next summer. Investigate past history if this is suspected (i.e. if vegetation is species-poor and made up of quick-growing ephemerals but the site has not recently been dredged).

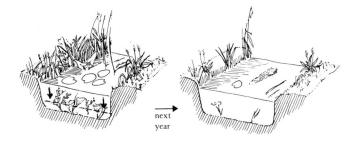

16. Summer drying

If channels are dry for part or all of the summer then this is not a source of damage in the correct sense, as the vegetation of these channels is described, in Chapter 3, as summer-dry.

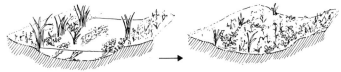

17. Fierce spates

Properly classed here are the very rare (e.g. once in 25–100 years) spates causing severe damage to vegetation in rivers normally able to support considerable vegetation. However, sites may also be placed here because the flow regime, as in parts of Devon, is fiercer than expected from the topography (see above and Chapter 2).

18. Very swift normal flow

This is not a source of damage in the correct sense, but sites may be placed here because the flow regime, as in parts of Devon, is fiercer than expected from the topography (see above and Chapter 2).

19. Regulation of flow (through reservoirs, water transfers, gates, etc.) leading to more irregular discharges

Flows leading to, in effect, more spatey conditions move a stream, vegetationally, into a more mountainous category. Regulation of flow leading to more stable discharges increases vegetation (see above).

20. Any other obvious source of damage

Chapter 6 can be consulted for further information on points 1—8, 12, and by implication or to some extent, for 9 and 14—19; and see Chapters 2 and 3 and the early part of this chapter, for points 13 and 17—19.

Estimating pollution index

1. If there is no physical cause for damage to vegetation, then the pollution index letter is that of the damage rating, i.e. $a = A$, $c = C$, etc.
2. If one or more of the physical sources of damage is severe, and influences the whole reach, plants cannot be used to diagnose pollution and the index is U (unclassable).
3. If one or more of the physical damage factors is present and affects the whole reach but is mild, then plants can give some guide to the pollution index. If the damage rating is, say, d, that part of the damage attributable to pollution must be less than this, giving an index of $C+$ (C, B or A). If there is clearly a great deal of non-pollution damage, a rating of d has an index of $B+$ (B or A). Other nearby sites should be examined; a more accurate index can be obtained if these sites are physically undamaged but otherwise comparable.
4. If a physical damage factor affects only part of a reach (e.g. shade, cattle trampling, bridge turbulence), then the damaged part should be ignored and the index be derived from the undamaged part. This is possible where the undamaged

part is large enough to exhibit the full plant community (usually between one third and two thirds of the total site; this varies with the vegetation type, and experience is needed to judge the appropriate proportion).

Interpreting damage rating

It is sometimes possible to find the source of damage, or at least to narrow the possibilities, from the way in which the damage rating is made up. The ways in which different sources of damage affect components of the damage rating is shown in Table 4.3.

Irrigation

Water for irrigation should always be taken from streams with a pollution index of *A*. However, if an *A* stream is known to receive effluents, tests should be done using the stream water on the actual crops which need irrigation. Different plant species differ in sensitivity to different chemicals, and something completely innocuous to, say, tough plants like *Nuphar lutea* and *Schoenoplectus lacustris* may harm a particular crop.

Testing is especially important when water transfers are involved.

Table 4.3. *The effects of different types of damage on the components of the damage rating*

Type of damage	Diversity	Cover	Colour band*	% tolerant*	Blanket weed	*Potamogeton pectinatus*	Notes
Dredging	+	+	?	?	+	—	
Cutting	(+)	+	?	?	(+)	—	
Shade	(+)	+	—	—	—	—	
Herbicides on emergents	—	?	?	—	?	—	
Herbicides in channel	+	+	+	+	+	—	
Boats	(+)	+	?	+	—	—	Delicate submergents lost first, tough emergents last
Trampling, etc.	(+)	+	?	—	?	—	
Unstable sediment	(+)	+	—	(+)	?	—	Channel species lost first, leaving undue proportion of edge emergents
Drought	+	(+)	(+)	(+)	+	—	Land spp. as in % tolerant column. Loss of submergents which, in slow flows, moves colour towards *blue*
Storm flows and equivalent	(+)	+	?	—	?	—	
Eutrophication	—	—	+	?	?	—	
Turbidity	(+)	(+)	—	(+)	—	—	Submergents lost before emergents
Salt	+	(+)	—	—	?	+	
Town and industrial effluent	+	+	?	+	+	+	Last two columns irregularly affected, but can be markedly so

+, usually affected.
(+), usually affected if severe.
?, sometimes affected.
*These two parts of the damage rating have provision for scoring if all vegetation is lost. This is not included in the table.

5 Dykes, drains and canals

INTRODUCTION

The man-made drainage channels of alluvial plains have different names in different parts of Britain, but here we will adopt the names used in the Fenland, the largest such plain in the country. Small channels between fields are called dykes, and the larger ones which collect dyke water and carry it to the river are drains. (Channels used for navigation are called lodes, but as regards vegetation they can be included with drains.) Other common names are rhynes, in the west of England and South Wales (pronounced, and sometimes spelt, reens), and in the south of England, sewers. The other main type of flat man-made channel is the canal, constructed for navigation. Canals usually lie above ground water level and so must have non-porous beds and sides, usually of clay, to prevent the water draining away.

All these man-made channels have negligible flow, and require constant maintenance and management. Canals have low banks, because water level never fluctuates much, while dykes and drains may have high or low banks (i.e. with the water well below, or near, ground level; banks are high in the more-drained areas). The banks are normally very steep or vertical at and below water level (see Figs below). None have much water movement, even during storms (canals have storm run-off channels, and though the water level of dykes and drains may rise considerably after rain, there is very little force of flow).

The survey methods are as described in Chapter 1, but there are two modifications. First, the flow type encountered is usually negligible and never rapid. Secondly, the smaller channels can be easily recorded from the sides, as in the narrow dykes it is possible to look down on the whole channel from the banks.

DYKES AND DRAINS

Introduction

The main alluvial plains are the Fenland, the area around and near the Norfolk Broads, Romney Marsh, Pevensey Levels, Somerset Levels, Gwent Levels, and the Trent Flood Plain. The vegetation types described here are those typical of these large areas and of farmland (arable or managed grassland). There are in addition many other smaller plains with dykes, also on silt, fen peat or clay. These are often either affected by water from the land above (see Chapter 3, Alluvial streams), or too shallow to allow a good water-supported vegetation, and they tend to be species-poor. Such dykes come closer to those described below in types 1 and 2 than in 3, and care should be taken in assessing damage ratings unless a species-rich water-supported vegetation is present in at least a few of the channels. Ditches in flood plains with acid bog peat are not included. Their vegetation is between that described below and that of blanket bog streams (see Chapter 3).

Areas well-protected from intensive farming, e.g. some Nature Reserves, have greater diversity, often especially of emergents, and have rather different species assemblages. Details of one such area are given in M. Grose and D. Allen (1978), *Water Plants of the Ouse Washes*, Royal Society for the Protection of Birds.

Channels of types 1 and 2 below cannot be given a damage rating as shading and/or water regime are too extreme, but some guidelines may be obtainable from the text and the charts for type 3 channels (Tables 5.1 and 5.2).

Type 1. Channels dominated by tall emergents (Fig. 5.1)

These are usually the smaller dykes and ditches, and the shallower disused larger channels that have silted up. All channels up to *c.* 0.5 m deep, and most of those to *c.* 1 m deep, will be dominated by tall monocotyledons if they are left unmanaged, while those dry for much of the year may have tall dicotyledons as well as, or instead of, the tall monocotyledons.

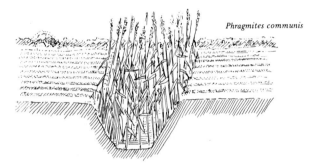

Phragmites communis

Fig. 5.1

Most common dominants:

Glyceria maxima, particularly in very shallow channels and outside eastern England

Phragmites communis, the most abundant British dominant

Species less frequently dominant, and more likely to dominate short stretches of the channel only:

Carex acutiformis, mainly southern

Scirpus maritimus, towards the coast

Carex riparia, mainly on fen peat

Sparganium erectum, not in drier channels

Phalaris arundinacea, most often in summer-dry dykes, and especially on clay

Typha spp., infrequent and scattered

Other potential dominants (e.g. *Carex pseudocyperus*) are local or rare in their distribution.

Type 2. Not as above, but too shallow to bear a species-rich water-supported vegetation (Fig. 5.2)

A species-rich flora of water-supported species can occur in only 25 cm of water if this level is stable, but depth usually varies through the year and a normal summer depth of at least 50 cm is generally needed to ensure good vegetation.

Alisma plantago-aquatica

Juncus effusus

Callitriche spp.

Fig. 5.2

Shallow channels without tall monocotyledons may have:

Drier: *Alisma* spp.   *Rumex* spp.

Juncus spp.   Small grasses

Wetter: *Callitriche* spp.   Small grasses

Lemna minor agg.   Blanket weed

Some species from Table 5.1 may also be present, and if there are enough of these some diagnosis and assessment may be possible for rock type or damage.

Type 3. Channels potentially with a good water-supported vegetation (Figs. 5.3 and 5.4)

These are channels not covered with tall emergents, and at least 50 cm deep (or 25 cm, if they have a very stable water level). They are (0.5–)2–20(+) m wide, with very steep to vertical sides so that in deeper channels emergents are confined by the depth of the water to narrow bands at the sides of the channel. In other channels emergents are kept sparse by management. Undamaged channels normally bear at least nine species per site and have at least 80% cover; this applies to both dykes and drains. As all channels are much-managed, 'undamaged' here means sites not cut for at least a month, not dredged for at least 2 years (or at least 4 months if in in a protected area, as defined above), and with cover and diversity left unaffected by any other damage factor (also see below). In practice, however, dykes (usually 2–4 m wide) and drains (usually 6–14 m wide) are likely to differ in vegetation. Drains are normally important for the large-scale movement of water in the plain, and therefore must be kept clear, while dykes are more likely to be allowed to bear at least some vegetation. In addition, drains, being usually deeper, are less likely to have emergents in the centre and more likely to have scour from wind and waves at the edges. They are also more likely to collect silt on the bed, and so to have a *redder* vegetation.

Table 5.1 shows the species that can be found in type 3 channels, together with some information on their habitats. The basis on which the species were grouped is described below.

Phragmites communis

Sagittaria sagittifolia
Callitricha platycarpa
Potamogeton natans
Elodea canadensis
Myriophyllum spicatum
Potamogeton pectinatus

Fig. 5.3

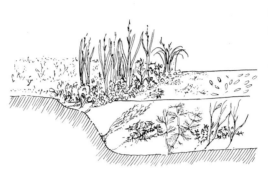

Glyceria maxima

Carex acutiformis
Veronica anagallis-aquatica agg.

Rorippa nasturtium-aquatica agg.
Veronica beccabunga
Myosotis scorpioides
Lemna minor agg.
Callitriche platycarpa
Ceratophyllum demersum
Myriophyllum spicatum
Elodea canadensis
Potamogeton natans

Fig. 5.4

Group 1. Tolerant species

These species are the most tolerant to management. *Callitriche* species are mainly found in water less than 75 cm deep. If these occur alone, no other habitat comment can be made.

Group 2. Widespread species

These species are widespread, geographically and ecologically, although fringing herbs are rare in drains. They are less tolerant to management than those of group 1. If enough of the species are present to place the site in one of the three subgroups (peat, silt or clay), a provisional diagnosis of soil type can be made. If, on further investigation, the soil type is found not to be that diagnosed, the cause should be sought — e.g. eutrophication of peat leading to a silt-type vegetation.

Group 3. Infrequent species

(NB Other, rarer species are also found in dykes.)

These species may be present alone, but they are normally accompanied by species from groups 1 and 2. As on the stream dial (Chapter 2), they are arranged in order of nutrient status and colour-banded for easier identification. Only three colour bands are given, and the *turquoise* is equivalent to both the *blue* and the *green* mesotrophic bands of the stream dial.

Habitat identification is firstly, as for streams, by identifying the colour most appropriate to the species of the site (provided sufficient species are present to do this). Secondly, there are additional notes to the right of the colour band list. The first arrow, going the whole length of the list, shows that all colours can occur on fen peat, and the notes beside it apply to sites on fen peat. *Yellow* species are effectively restricted to fen peat. The second arrow marks the species characteristic of silt, and this again reaches to the end of the *red* band but starts with the *turquoise* band. In fact, the more nutrient-poor *turquoise* species are common on

Table 5.1. *Dykes and drains chart*

1. TOLERANT SPECIES	2. WIDESPREAD SPECIES	
Agrostis stolonifera and other small grasses	(*a*) Alphabetical list	(*b*) Habitat lists (alphabetical)
	Alisma plantago-aquatica agg.	
Callitriche spp. (particularly *C. platycarpa*)	*Apium nodiflorum*	(i) Peat
		Carex acutiformis agg.
Enteromorpha sp.	*Carex acutiformis* agg. (Fenland)	*Ceratophyllum demersum* agg.
Blanket weed	*Ceratophyllum demersum* agg. (species-rich sites)	*Glyceria maxima*
If near coast:		*Lemna minor* agg.
	Elodea canadensis	*Nuphar lutea*
Potamogeton pectinatus	*Glyceria maxima*	*Phragmites communis*
Scirpus maritimus	*Lemna minor* agg.	*Sagittaria sagittifolia*
	Nuphar lutea	*Sparganium erectum*
	Phalaris arundinacea	(ii) Silt
	Phragmites communis	*Alisma plantago-aquatica* agg.
	Potamogeton natans	(*Apium nodiflorum*)
	Potamogeton pectinatus	*Carex acutiformis* agg.
	(*Potamogeton perfoliatus*)	*Ceratophyllum demersum* agg.
	Rorippa nasturtium-aquaticum agg.	*Glyceria maxima*
	Sagittaria sagittifolia (Western)	*Lemna minor* agg.
	Sparganium erectum	*Phalaris arundinacea*
		Phragmites communis
	The three main habitats do, however, *tend* to have somewhat different species groupings	*Potamogeton pectinatus*
	Additional habitat notes are given in brackets. These are tendencies, not exclusive preferences (e.g. 'Fenland' means more in the Fenland than in the other regions)	*Potamogeton perfoliatus*
		(*Rorippa nasturtium-aquaticum* agg. near lowland)
	Hydrocharis morsus-ranae and the other *Lemna* spp. are locally very abundant, but appear only in group 3	*Sagittaria sagittifolia*
		Sparganium erectum
		(iii) Clay
		Apium nodiflorum
		Carex acutiformis
		Glyceria maxima
		Lemna minor agg.
		Phalaris arundinacea
		Phragmites communis
		Potamogeton natans
		Potamogeton pectinatus
		Rorippa nasturtium-aquaticum agg.

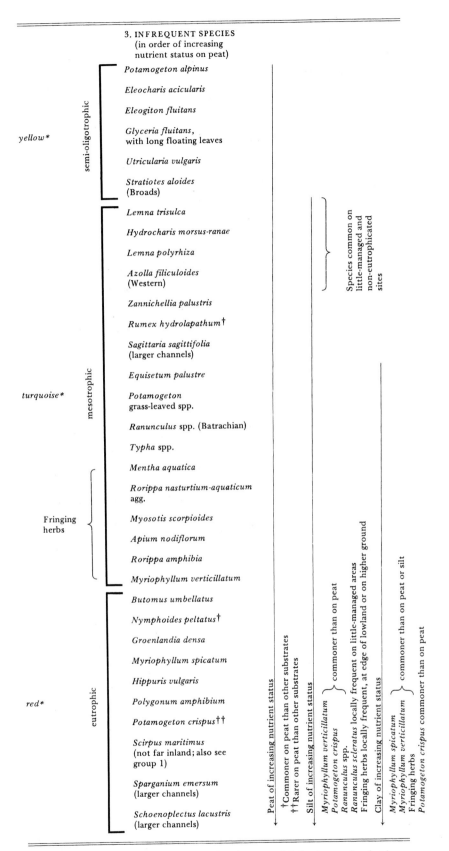

3. INFREQUENT SPECIES
(in order of increasing
nutrient status on peat)

yellow* semi-oligotrophic

Potamogeton alpinus

Eleocharis acicularis

Eleogiton fluitans

Glyceria fluitans,
with long floating leaves

Utricularia vulgaris

Stratiotes aloides
(Broads)

turquoise* mesotrophic

Lemna trisulca

Hydrocharis morsus-ranae

Lemna polyrhiza

Azolla filiculoides
(Western)

Zannichellia palustris

Rumex hydrolapathum†

Sagittaria sagittifolia
(larger channels)

Equisetum palustre

Potamogeton
grass-leaved spp.

Ranunculus spp. (Batrachian)

Typha spp.

Fringing herbs {

Mentha aquatica

Rorippa nasturtium-aquaticum
agg.

Myosotis scorpioides

Apium nodiflorum

Rorippa amphibia

Myriophyllum verticillatum

red* eutrophic

Butomus umbellatus

Nymphoides peltatus†

Groenlandia densa

Myriophyllum spicatum

Hippuris vulgaris

Polygonum amphibium

Potamogeton crispus††

Scirpus maritimus
(not far inland; also see
group 1)

Sparganium emersum
(larger channels)

Schoenoplectus lacustris
(larger channels)

Species common on
little-managed and
non-eutrophicated
sites

Peat of increasing nutrient status

†Commoner on peat than other substrates
††Rarer on peat than other substrates

Silt of increasing nutrient status

Myriophyllum verticillatum
Potamogeton crispus
Ranunculus spp. } commoner than on peat
Ranunculus sceleratus locally frequent on little-managed areas
Fringing herbs locally frequent, at edge of lowland or on higher ground

Clay of increasing nutrient status

Myriophyllum spicatum
Myriophyllum verticillatum } commoner than on peat or silt
Fringing herbs
Potamogeton crispus commoner than on peat

If additional species are present which are listed on the stream dial, the stream colour band
usually applies

*Readers should colour these bands for themselves, in the way that the stream dial has been
coloured.

silt only in little-managed (i.e. with some dredging and cutting, and little spraying) and non-eutrophicated areas. A few habitat notes beside the arrow record other differences between silt and peat vegetation. The third arrow marks the species characteristic of dykes on clay, which are the most nutrient-rich, and this arrow takes in only the lower half of the *turquoise* band in addition to the *red* one. There are again some habitat notes beside the arrow.

In contrast to streams, dykes and drains often show great variation in species composition along short stretches of channel. The communities should, however, fall within the same colour bands or species groups. (As in streams, it is quite possible to have, for example, a *turquoise* community with one or two *red* or *yellow* species present also.)

There is some geographical variation, e.g.

Azolla filiculoides, south and south-west England
Nymphoides peltata, eastern England
Stratiotes aloides, East Anglia

Seasonal variations, within the summer season, are greater than in streams. Some potential dominants are usually at their optimum before mid-June (e.g. *Callitriche* spp., *Hottonia palustris*), while others may not even be up until this time (e.g. some *Potamogeton* spp., *Sagittaria sagittifolia*).

Dykes and drains rating

The rating, like that of streams, is from *a* for all right to *h* for horrible. However, at present only a five-point rating can be given (*a, b, d, f, h*), and no pollution index. Sites which cannot be rated (those with tall emergents or too-shallow water (i.e. those in types (1) and (2) above) are termed *u*, for unratable. It is hoped to improve this in the future to an eight-point scale (and so the rating is now given from *a* to *h*, to correspond with the stream rating), but since most of the damage to dyke vegetation comes from herbicides and other management techniques it is unlikely that, in the near future, an index can be given to separate the effects of herbicides etc. and of poisonous effluents.

The rating does not assess eutrophication (unless this is accompanied by other damage factors). Eutrophication is likely: on peat, at any *red* site, or at lower *turquoise* sites on dykes; on silt, at *red* sites except on large drains, and at lower *turquoise* sites on smaller dykes; on clay, at *red* sites on dykes. Such sites should be checked for the influence of fertilisers etc., or of incoming water bringing silt from higher ground.

Table 5.2 shows how ratings for dykes and drains are assessed. Dykes in protected areas are likely to be 27+ and be rated a_1. Good sites in farmland are more likely to be 20—26 and be rated a_2. This is due partly to the farming practices and partly to the typically higher water level in the protected areas: low water levels mean steeper banks at the water's edge and so less available habitat for emergents, while gently sloping banks with the water near the top, and grazed edges, both increase the available habitat for edge aquatics (see Figs.).

Possible sources of damage include:
1. Recent dredging (within *c*. 2 years, or, in protected areas, *c*. 4 months)
2. Recent cutting (within *c*. 1 month)
3. Herbicides
4. Recent drying or shallowing
5. Excessive and prolonged increase in water level
6. Trampling, paddling, cattle disturbance, etc.

Table 5.2. *Rating for dykes and drains*

Add score for species present	Add species diversity allowance		Add cover value for all species (not including tall edge plants)		Damage-rating	
	No. of species present	Add		Add	Total	Rating
Tolerant species score 1 / Widespread species score 2 / Other species* score 3	0–3	0			0–1	h
	4–6	1	Cover 50%	0	2–5	f
	7–9	3	Cover 50–70%	1	6–10	d
	10+	5	Cover 75–100%	2	11–19	b
					20+	a

If shallow or covered with tall emergents, describe separately (see text)

20–26 rate a_2
27+ rate a_1

*Species other than those of groups 1 and 2, not necessarily those listed in group 3 of Table 5.1.

7. Boats (in drains or large dykes)
Some guide to the source of the damage can be obtained from the type of effect it has (Table 5.3).

Potamogeton pectinatus is usually characteristic of recent dredging or of salt.

Turbidity can be due to: clean but turbid stream water entering the dyke system (e.g. Pevensey Levels); silting from arable land; phytoplankton; herbicides; boats, in larger channels and in dykes connected to these; or various effluents. Clean turbid water affects biomass but not diversity or cover, and sites are therefore rated as undamaged (*a*). Damaged turbid sites also have toxins in the water, or disturbance from boats, etc.

CANALS

Introduction

Canals occur over much of England, into the east of Wales, and across the central lowlands of Scotland. The canals proper are channels specifically made for navigation, but there are also a number of rivers in the lowlands which have been canalised. The latter are considered both here and under slow streams of size (iv) in Chapter 3.

Three groups of canals lie outside the general pattern and are not further discussed here:
1. Disused and silted-up canals, which are seldom or never flooded, and are usually dominated by *Glyceria maxima*, less often by *Phragmites communis*.
2. Ship canals, which are particularly large, turbid and carry large ships. These have negligible vegetation.

Table 5.3. *The effect of different types of damage on the components of the damage rating*

Type of damage	Part of rating affected				Notes
	Diversity loss	Cover loss	Tolerant spp. important	Infrequent spp. important	
Dredging	(+)	+	—	+	
Cutting	(+)	+	—	+	
Herbicides	+	+	+	—	Yellowing of plants
Drying	+	?	—	—	Loss of water-supported spp., particularly slow-growing spp.
Flooding	+	(+)	—	—	Loss of shallow-water spp. (see Chapter 6)
Trampling, etc.	+	+	—	+	
Boats	+	+	(+)	—	Loss of water-supported spp.

+, much affected
(+), less affected
?, possibly affected

3. The Caledonian Canal, which has a stone and gravel bed, deep water, and negligible vegetation.

Canal vegetation

Canals have basically the same cross-section as drains, but were originally constructed with large coping stones on the towpath side. Often the stones have fallen in, forming a ledge just below water level, on which plants can grow. These may be emergents, or water-supported species, depending on the habitat. Canals are often turbid. (Figs. 5.5, 5.6).

Table 5.4 lists the species that can be found in canals, grouped according to their tolerance. The basis on which the species were grouped is described below.

Group 1. Tolerant species

These species are the most tolerant to boat disturbance and pollution (the pollution being more from effluents than from the boats). If these species occur alone, no other habitat comment can be made.

Group 2. Semi-tolerant species

These are less tolerant than those of group 1, but still occur under conditions of severe stress.

Group 3. Colour-banded species

(NB Other, rarer species are also found in canals.)

These species may be present alone, but are often accompanied by species from groups 1 and 2. As on the stream dial (Chapter 2) and the dykes and drains chart (Table 5.1) they are arranged in order of nutrient status, and colour-banded for easier identification. Only three colour bands are given, and the *turquoise* is, like that in Table 5.1, equivalent to both the *blue* and the *green* mesotrophic bands of the stream dial.

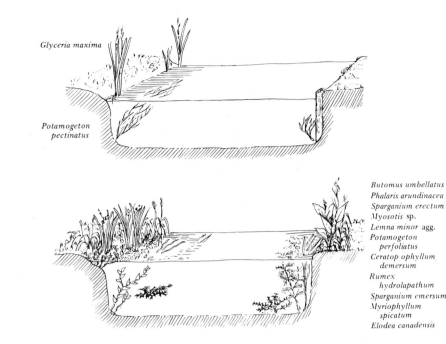

Fig. 5.5

Fig. 5.6

Glyceria maxima

Potamogeton
pectinatus

Butomus umbellatus
Phalaris arundinacea
Sparganium erectum
Myosotis sp.
Lemna minor agg.
Potamogeton
    perfoliatus
Ceratop ophyllum
    demersum
Rumex
    hydrolapathum
Sparganium emersum
Myriophyllum
    spicatum
Elodea canadensis

Habitat identification is by identifying the colour band most appropriate to the species of the site (provided sufficient species are present to do this). Most English canals are on eutrophic clay with *red—purple* colour bands. On alluvial silt and on sandstone the colour is *turquoise—purple* to *purple*, on peat *turquoise*, and on Resistant rock and chalk *turquoise—purple*. However, as in streams, the rock type at the source of the water is important where this differs from the rock type of the site (see the sections on mixed catchments in Chapter 3). Canals which are *redder* than this should be checked for eutrophication. However, several *red* species also occur in group 2, as semi-tolerant species, and if they alone are present the site is not necessarily eutrophicated.

Canalised rivers are most often on clay. Vegetation and flow regime are similar to, but in between those of, slow clay rivers and canals on clay — having *red* and *purple* species.

As the boat density on canals increases, the vegetation is lost, in the following order:

Water-supported species in the centre of the channel
The more sensitive water-supported species in the bands at the sides
The side bands, leaving scattered water-supported species there, but still emergents on the banks (which can protect the banks from erosion)
Most of this vegetation, leaving occasional emergents and rare water-supported species
The remaining vegetation

If canals have little or partial dredging, and become silted, particularly on the non-towpath side, emergents (both tall ones and fringing herbs) can grow very well here. On the towpath side boats tend to keep the canal clearer, especially where they can tie up. However, the non-towpath side is often shaded, and this decreases the vegetation there.

Locally good vegetation often develops in winding holes (wide pools out into, usually, the non-towpath side, where boats can turn round). These are often poorly maintained, and emergents in particular can grow well. Locally good vegetation can also occur in disused locks, where one of a pair of locks has not been kept repaired; in side ponds (where water can be transferred from a lock to a side

105

Table 5.4. *Canal chart*

1. TOLERANT SPECIES	2. SEMI–TOLERANT SPECIES
Acorus calamus	*Butomus umbellatus*
Glyceria maxima	*Carex acutiformis*
Lemna minor agg.	*Nuphar lutea*
Potamogeton pectinatus	*Phragmites communis*
Sparganium erectum	*Rumex hydrolapathum*
Blanket weed	*Sparganium emersum*
(*Agrostis stolonifera* and other small grasses, but infrequent)	*Enteromorpha* sp.

3. COLOUR-BANDED SPECIES

Utricularia vulgaris	
Hydrocotyle vulgaris	
Hydrocharis morsus-ranae	
Berula erecta	
Lemna trisulca	
Mentha aquatica	turquoise*
Lemna polyrhiza	
Ranunculus spp. (short-leaved)	
Callitriche spp.	
Potamogeton natans	
Mosses	
Grass-leaved *Potamogeton* spp.	
Elodea canadensis	
Iris pseudacorus	
Potamogeton perfoliatus	
Ceratophyllum demersum	
Potamogeton lucens and hybrids	
Juncus effusus	
Phalaris arundinacea	purple*
Myosotis scorpioides	
Typha spp.	
Polygonum amphibium	
Alisma plantago-aquatica	
Myriophyllum spicatum	
Potamogeton crispus	
Epilobium hirsutum	
Sagittaria sagittifolia	
Enteromorpha sp.†	
Butomus ambellatus†	red*
Sparganium emersum†	
(*Schoenoplectus lacustris*)	
Nuphar lutea†	

If additional species are present which are listed on the stream dial, the stream colour band usually applies.

*Readers should colour these bands for themselves, in the way the stream dial has been coloured.
†Also listed in group 2.

pond and then to the next lock); and in places where the canal is wider than usual and boats seldom pass near the non-towpath side.

Canals rating

The rating, like that of dykes and drains, is from *a* for all right to *h* for horrible, and is, at present, only on a five-point scale (*a*, *b*, *d*, *f*, *h*), without a pollution index. It may be possible to develop an eight-point rating soon, but since most damage to canal vegetation is from boats and effluents, boats being the more important factor, it is unlikely that, in the near future, an index can be given to separate the effects of these.

There is no genuinely undamaged vegetation, as management is needed to keep this aquatic habitat in being. Canals here termed undamaged are sites where diversity, cover and quality of vegetation are not obviously harmed.

The rating does not assess eutrophication (unless this is accompanied by other damage factors). The expected colour bands for canals on different rock types are given above, and *redder* vegetation should be investigated for eutrophication.

Table 5.5 shows how ratings for canals are assessed. Canals with the highest ratings of 30+ are usually those with a habitat which is stable, unpolluted, and little-harmed by boats.

Most damage is from boats and pollution but other possible sources include recent dredging (within *c.* 2 years), recent cutting (within *c.* 1 month), herbicides, anglers (clearing patches at the sides) and shading.

As in dykes and drains, clean turbid water affects biomass but not diversity or cover, and sites are therefore rated as undamaged (*a*). Damaged turbid sites are those disturbed by boats or poisoned by effluents, etc.

Table 5.5. *Rating for canals*

Add score for species present	Add species diversity allowance		Add cover value of water-supported species within band at sides		Rating	
	No. of species present	Add		Add	Total	Rating
Tolerant species score 1; If all species tolerant, subtract 1 / Semi-tolerant species score 2 / Other species* score 3	0–2	0			0–2	*h*
			Less than 5%	−1	Including negative rating	
	3–5	1			3–5	*f*
	6–8	3	Sparse 10–25%	+1	6–9	*d*
			Dense over 30%	+3	10–19	*b*
	9+	5			20+	*a*

If silted up and covered with tall emergents, (usually *Glyceria maxima*) describe separately (see text)

Within *a*, if total cover across channel is 60+%, and rating 30+, rate a_1, if cover less than 60% and/or rating 20–29, rate a_2.

*Species other than those of groups 1 and 2, not necessarily those listed in group 3 of Table 5.4. Canal species listed in both 2 and 3 are scored as semi-tolerant.

6 Effects of man's activities

INTERFERENCES ASSOCIATED WITH CHANNEL MAINTENANCE

1. Recommended vegetation

Plants can obstruct water movement, boats, anglers, sluices, etc. They also stabilise banks and channels, provide habitats and food for animals, aerate and purify water, are pleasing to look at, etc. The optimum vegetation for a watercourse is thus that which produces the most desirable and least undesirable effects. Recommendations for the optimum vegetation for different types of streams are given below.

Hill streams

In the hills, vegetation can be removed by spate flows, and therefore does not create serious flood hazards. Channels can usually be up to 75% full, except where clearer water is needed for gauges, anglers, etc.

In the flood plains, tall monocotyledons on the banks prevent erosion, and up to 50% total vegetation is acceptable in channels with frequent spates.

Lowland streams: soft limestone

Upper brooks can bear dense vegetation (and are suitable for conservation). Medium streams should be c. 25% full, as this is adequate for both flood prevention and conservation. Where it is possible to favour one species at the expense of another, *Ranunculus* should be decreased, as short species create less of a flood hazard. Trees, particularly alders, can be planted on the sunnier bank, to shade the stream. The same applies to large streams, except that tall monocotyledons on the banks should be encouraged, to protect these banks, particularly in flood plains.

Lowland streams: clay

Upper brooks should have shading trees or hedges on those flatter, smaller reaches where tall monocotyledons can choke the channel. Small brooks should also be shaded if tall dicotyledons grow thickly enough on the banks to create flood hazards there. Large streams should have (narrow) fringes of tall monocotyledons to stabilise the banks. In general, the physical characteristics of the channels prevent the safety limit — 25% full — from being exceeded.

Lowland streams: soft sandstone

Comments vary between those for clay and those for soft limestone streams, depending on type.

Channel maintenance

Canals

Tall monocotyledons on the banks prevent erosion and should be encouraged, even to the extent of specially protecting them against boats and anglers (if the cost of such protection is less than that of piling the sides). The ledge (made by the fallen coping stones) should be preserved, as this can bear good vegetation in a habitat partly protected from boats. When boat density is high enough to remove vegetation from the channel centre, weed control should be stopped.

Dykes and drains

The main waterways must be kept free of plants which could cause obstructions. In wide channels a narrow belt of tall monocotyledons at the sides may be allowed, as may short grasses (or trees) on the banks. Channels moving little water (at the head of a drainage system) may develop thick vegetation, and be suitable for conservation. Narrow channels moving water should have short grasses on the banks (to prevent flood hazards from the banks) or be shaded by trees.

2. Flood hazards from river plants

The risk of flooding varies between individual streams, and vegetation and history should be checked before management is planned. The typical pattern of flood hazards is:

High	*Medium*	*Low*	*Negligible*
Chalk	Chalk/clay	Most clay	Very mountainous
Dykes	Some upper clay	Some little-used canals	Mountain
Drains	Some soft sandstone Rarely, canals	Upland sandstone Some upland limestone	Upland resistant Most canals

3. Danger from flood hazards

The situations where weed control is necessary can be ascertained by combining the information in the table here with that above. The typical pattern of danger from flood hazards is:

High	*Medium*	*Low*	*Negligible*
Built-up areas Large plains with arable	Arable by rivers Large plains with grassland	Grassland by rivers	Mountain and moorland with little valley, or with little economic value to the valley (and few animals)

6 Effects of man's activities, etc.

4. Dredging

Usual frequency of dredging

2—5 years	5—12 years	12+ years
Dykes near edge of lowland	Dykes	Faster clay streams
Other shallow dykes used for much water movement	Drains	Lowland sandstone streams
	Slow- (and some moderate-) flow clay streams	Lowland chalk-mix streams
	Some slow mixed-clay streams	Upland sandstone streams (rarely)

Usual time for recovery after dredging

Time for recovery of vegetation to its former stability, and to an equivalent quantity (or to a lesser quantity if the vegetation was formerly choking the water-course), is shown in Table 6.1.

Table 6.1. *The time taken for vegetation to recover after different types of dredging*

	Years to recovery
1. Shallow dredge, not breaking hard bed*	1 and less
2. Deeper dredge not breaking hard bed (and excluding 3—6 below)*	3(—2)
3. Dredge breaking hard bed in non-silting stream	6+
4. Dredge breaking hard bed in silting stream†	2—6
5. Deepening channel sufficiently to decrease substantially the light reaching the bed†	4—10+
6. Altering flow so that a previously swift reach becomes deep and slow	Until silting permits invasion of slow-flow silt-growing species

Thorough dredging, removing plant propagules, leads to longer recovery times. Shorter recovery times occur in special circumstances

*Good for conservation

†Good for conservation in drains, etc. IF the alternative way of preventing flood hazards is aquatic herbicides.

Recognition of recently dredged sites

Very recent:
>   Bare soil (not rock or hardened mud) on banks
>   Channel usually empty
>   Blanket weed unduly luxuriant
>   Dredged spoil on top of bank

Less recent:
>   Dredged spoil on top of bank
>   Blanket weed unduly luxuriant
>   Channel intermittently poor in vegetation (dredging is usually done in short stretches)
>   *Sparganium erectum* sparse, curled and short
>   *Juncus effusus* on bank (in non-moorland areas)

More distant:
>   (Differences in bulk of vegetation along a uniform reach, in a stream likely to be dredged)

Juncus effusus on bank (in non-moorland areas)

5. Cutting

For methods and equipment see T.O. Robson (1974), 'Mechanical control', in *Aquatic Vegetation and its Use and Control*, ed. D.S. Mitchell, UNESCO, Paris.

Usual frequency of cutting
(when not combined with dredging and herbicides)
In channels:

2+ per summer	*1 per summer*	*c. 1 per two summers*
Soft limestone streams of sizes (iii), large (ii), shallow (iv)	Some soft limestone streams	Some sandstone streams
Dykes and drains with much water transport (or angling)	Dykes and drains with considerable water transport (or angling)	Some lowland clay streams
	Flatter clay brooks with thick tall monocotyledons	Where there is explosive growth in shallow channels after dredging
	Places where clumps of tall monocotyledons create hazards	Locally elsewhere
	Some slow clay size (iii) streams	
	Near gauges in hill streams	
	Ornamental reaches by gardens	
	Where needed by anglers, etc.	

On banks:
Trees should be maintained so as to prevent the danger of fallen wood impeding flow, and any fallen wood should be removed.
Herbaceous vegetation (and small woody plants) should be:

Summer-cut	*Winter-cut*
Paths on banks (2 per summer if little-trampled, none if much trampled)	Narrow brooks and dykes with high banks where tall vegetation creates hazards in winter storms
Dykes and narrower drains with high banks and tall vegetation	

One factor against cutting as a means of controlling vegetation is that it is labour-intensive and so, in Britain, costly. There are also situations where it should be carefully controlled or avoided altogether. For example, if the animal life is rich or rare tall vegetation should be cut in short stretches. Tall plants used for nesting (e.g. reeds by reed-bunting) should not be cut during the nesting season. Cutting is unnecessary where species are non-invasive and sparse. *Ranunculus*, the most troublesome plant of flowing waters, grows more if it is cut. If, therefore, the stream can safely be left uncut, this will decrease the management needed and there will be less bulk of plants the next year (F.H. Dawson, personal communication).

With these exceptions, cutting is recommended for conservation, and, in many dykes and drains is the only means of preserving both water transport and a water-supported vegetation. More cutting is needed in hot summers. On banks, one to two cuts per year leave tall rough herbaceous vegetation dominant, while more frequent cuts increase the tendency towards a grass sward.

Usual time for recovery after cutting

Channel species:
> Usually 2—4 weeks in early- or mid-summer to replace cover, longer for biomass. Cutting in late summer leads to little re-growth of strongly seasonal species (e.g. *Potamogeton pectinatus*) though non-seasonal ones (e.g. *Callitriche* spp.), in suitable temperatures, are unaffected by the date of cutting.

Bank species:
> Over a month for non-seasonal species and species cut in early spring, but there is little recovery in late summer for strongly seasonal species (e.g. *Phragmites communis*), and their crop is reduced in the next year.

Recognition of recently cut sites

> Cut stumps of plants
> Decomposing vegetation on banks or cut fragments in water
> Too little vegetation for the type of channel, biomass unduly low for the cover
> (needs experience to judge)

6. Herbicides

Approved herbicides, their application and the expected results are described in T.O. Robson (1978), 'Recommendations for the control of aquatic weeds', in *Weed Control Handbook*, 8th edn, ed. J. Fryer and R. Makepeace. Blackwell, Oxford; and T.O. Robson (1973), *The Control of Aquatic Weeds, Bulletin 194*, Ministry of Agriculture, Fisheries and Food. Herbicides must be used in accordance with the Ministry of Agriculture, Fisheries and Foods (1979) *Code of Practice for the Use of Herbicides on Weeds in Watercourses and Lakes*.

At present (1979) herbicides may be used on banks, on local patches of emergents in streams, and in dykes, drains and canals whose water is not used for irrigation, etc.

In still or nearly still waters which are only intermittently used for irrigation, herbicides can be used. After application, the concentrations of herbicides and their residues decrease, until these are low enough for the water to be used with safety (see labels on tins of herbicides). Herbicides for water in streams are being developed.

Herbicides are more efficient than cutting, in that fewer applications are needed and a total kill can be achieved easily.

Herbicides used

In channels:
> Chlorthiamid   Diquat
> Dichlobenil   Terbutryne

can be used to kill plant parts within the water, and usually affect thin delicate parts more than tough thick ones (see references above for further details). Spraying should be done at the start of the growing season, to prevent growth. Spraying thick mature vegetation can lead to decomposing vegetation in the water, and consequent deoxygenation and death of animals, especially fish.

Glyphosate can be used on the tough floating leaves of *Nuphar lutea*. Glyphosate has a specific action, killing only the actual plants which are sprayed. (The glyphosate is moved within the plant, so it is not just the sprayed parts which are harmed, but the whole of the plants.)

The others, being non-specific, can rarely be recommended for conservation. Spraying is cheaper than cutting, and so aquatic herbicides are often the only practical means of weed control, and flood prevention. However, herbicides in water can spread for considerable distances from the point of application, and can, of course, damage vegetation over this distance. An unsuitable choice of the type of herbicide and the habitat in which it is used can lead to the replacement of macrophytic vegetation by blanket weed, with no gain to water transport and a loss to conservation.

If physical factors (e.g. depth) prevent vegetation from growing enough to create flood hazards, sparying is unnecessary.

On banks, and emergents in channels:

Dalapon	Maleic hydrazide
Dalapon—paraquat	Maleic hydrazide—2,4-D

The first three and glyphosate are used to kill or diminish tall monocotyledons (e.g. *Phragmites communis*, *Typha* spp.), and the last two for converting tall bank vegetation to a short grass sward.

Tall vegetation (if sprayed when grown) should be treated in short stretches, if the animal life is rich or rare, to allow animals to move to cover. Tall plants used for nesting (e.g. reeds by reed-bunting) should not be sprayed during the nesting season. Spraying is unnecessary where species are non-invasive and sparse.

With the above provisos, spraying of tall monocotyledons is satisfactory for conservation. It is now possible to produce a population of sparse tall monocotyledons that is very suitable, in shallow dykes carrying little water, for both flood prevention and conservation. Removing tall monocotyledons from dykes able to support a good water-supported vegetation can aid conservation if the latter can be allowed to develop.

Usual time for recovery after spraying

Channel species:
   1—4+ years, depending on the availability of fragments and propagules for re-invasion. The replacement vegetation may be different from the original vegetation.
Emergents and bank vegetation:
   3+ years, if the vegetation was totally removed.

Recognition of recently sprayed sites

In the channel:
   Dead stumps with blanket weed
   Yellow and dying vegetation
   Turbid empty channels with no obvious pollution
   Diversity and cover decreased with no other good reason — check
Emergents and bank vegetation:
   Dead stumps

Yellow and dying vegetation
Sparse short shoots where tall dense ones are expected
Grass swards which are not grazed

7. Shading

The use and cost-effectiveness of shading by trees for weed control are described in F.H. Dawson and U. Kern-Hansen (1978), 'Aquatic weed management in natural streams: the effect of shade by the marginal vegetation', *Vehr. Internat. Verein. Limnol.* 20, 1451–6; and W. Lohmeyer and A. Krause (1975), 'Ueber die Auswirkungen des Geholbewichses an kleinen Wasserläufen des Munsterlandes auf die Vegetation im Wasser und an den Böschungen im Hinblick auf die Unterhaltung der Gewasser', *Schweiz. Reihe Veg.* 9, 105pp.

For many centuries shading of brooks by hedges has been recognised as effective in weed control. Pilot tests on planting trees, particularly alders, have proved effective in Britain and Germany. Trees are planted so that they will form a closed canopy with light shade, and are effective after *c.* 4 years. If planted only on the sunny bank and on channels up to *c.* 8 m wide, the line of trees both provides satisfactory weed control and allows access by heavy machinery. The sparse, separated trees so common on British stream banks cast too little shade to prevent flood hazards.

It is not possible to have trees on both banks of much-silted channels requiring regular access by dredging equipment (note, however, that silting is increased by vegetation and so this point is not relevant when empty channels do not silt up). Trees on even one bank are not possible where trees themselves create a flood hazard.

The same effect is provided by shading from bridges, buildings, etc.

Species differ in their tolerance to shade, and, where light shading is involved, a canopy sufficient to reduce e.g. *Ranunculus* can leave e.g. *Sparganium emersum* unaffected (see S.M. Haslam (1978), *River Plants*, Cambridge University Press). In order of increasing tolerance to shade, some main dominants of British streams are ranked:

Ranunculus spp.

Rorippa nasturtium-aquaticum agg.

Apium nodiflorum,   Berula erecta,   Glyceria maxima,   Nuphar lutea,   Sagittaria sagittifolia

Certaophyllum demersum,   Elodea canadensis

Callitriche spp.,   *Phragmites communis*

Sparganium erectum

Sparganium emersum

Turbidity

Turbidity means shading by particles or coloured solutes within the water. It is frequently associated with pollution (e.g. effluents, herbicides), abrasion or unstable substrates and so it can be difficult to separate its effects from those of associated factors. Shading from turbidity can only affect plant parts in the water, and therefore generally harms submergents the most. In order of increasing tolerance to turbidity, some main dominants of British streams are ranked:

Elodea canadensis, *Ranunculus* spp.

Callitriche spp., *Myriophyllum spicatum*, *Potamogeton natans*, *Potamogeton pectinatus*, *Sparganium emersum*, *Sparganium erectum*

Ceratophyllum demersum, *Nuphar lutea*, *Sagittaria sagittifolia*, *Schoenoplectus lacustris*

8. Grazing

Grazing is very effective for keeping bank vegetation short. Grazed edges are found along most hill streams, some lowland ones, and also dykes and drains in some alluvial plains that are still (partly) under grass (e.g. Romney Marsh).

For grazing to be effective, animal access must be possible and the size of the animal must be appropriate to the stability of the bank. For example sheep can safely maintain many banks which would be seriously damaged by cattle or horses (though some banks are too unstable even for sheep).

9. Bank protection

Soil banks, unless protected or much compacted, can be eroded by currents or waves or, if they are steep, by soil falling. Compaction occurs in clay soils where flow for a long period has hardened, without destroying, the banks. This cannot be relied upon under other circumstances, and high banks of bare soil (except those bare for only a short period after dredging) should be considered unsafe.

Vegetation protecting banks can be:

(*a*) Trees and shrubs (preferably alders, as their roots grow deep: see the Lohmeyer and Krause reference in section 7 above). However, for reasons of land use, flood hazard or bank height it may not be possible to have trees on the bank.

(*b*) Grass (short swards). This is the usual vegetation near water level in hill streams, and is occasional elsewhere. It requires regular maintenance (by grazing or herbicides). Grass swards are very satisfactory in areas of high flood hazard, as, being short, they do not create a hazard on the bank and, being uniform, they are unlikely to be disrupted and eroded in storms. If water force is high, however, the grasses should be long-rooted species.

(*c*) Tall monocotyledons (e.g. *Phragmites communis*). These have deep rhizome networks, so provide excellent basic stabilisation. The uniformity of the stand means it is unlikely to be disrupted and eroded in storms. However, in narrow channels (particularly dykes) where tall vegetation on the banks itself creates a hazard, this type of vegetation can be dangerous.

(*d*) Tall dicotyledons and mixed herbaceous vegetation. This is the normal bank vegetation in most of the soft-rock and some of the upland areas of Britain. Necessarily, therefore, it is an adequate means of protecting the banks. It is, however, less good than the other possibilities above, since the plants are often short-rooted and form non-uniform stands, and so are liable to erosion in storm flows.

Narrow bands of tall monocotyledons should usually be encouraged on channel banks. Herbicides should be used to clear (completely remove) bank vegetation only after a pilot scheme has proved there is no danger to bank stability. Sand and silt banks (especially steep ones) are particularly vulnerable to erosion.

Other means of protection — normally much more expensive — are needed when vegetation has been lost, usually by excessive disturbance by boats, excessive use of herbicides, the joint effects of flow and pollution killing bank vegetation, or local extra stress (e.g. by bridges).

The remedies include:

Piling (typically in canals and canalised rivers)

Concreting or bricking the channel (often near bridges, but also elsewhere: see section 10 below)

Stone-walling the channel (more common outside Britain)

Formerly, mats of osiers or reeds were also used for bank protection

10. Concreting, bricking, etc.

Channels may be completely concreted or brick-lined when under severe stress, or may have just their banks protected when the stress is somewhat less. This is needed typically:

Where channels pass under roads etc.

In built-up areas where seepage or flooding would be dangerous

Near bridges, weirs, locks and other structures

Where a channel has been altered from its natural size and shape (and cannot retain the new pattern in the incident flow regime)

In grossly polluted channels where bank vegetation is killed and the banks erode

(In some complex irrigation systems with channels at different levels: more common outside Britain)

These channels range from those with near-natural substrates, able to maintain thick water-supported vegetation, to those with bare concrete and no vegetation. Bank plants are usually more affected than channel ones, as silt cannot properly accumulate on steep concrete slopes.

Channels which are concreted should be diagnosed as having either complete damage (*h*), or having damage of unknown extent. Experienced workers should be able to give some estimate of the latter.

11. Shaping channels

See section 4 on dredging for effects, recovery and recognition (and also read section 10 above).

INTERFERENCE ASSOCIATED WITH USES OF WATER AND WATERCOURSES

12. Altering (normal) flow type

(see section 13 below for storm flows)

The normal flow type can be altered by the following factors.

In long stretches:

Large reservoirs near source, with controlled discharges

Canalisation

Gates or other structures sufficiently close together to hold up (retain) the water all along the stretch

Water transfers

Abstraction (large quantities)

In short reaches:

Reservoirs, not as above

Abstraction, not as above

for this and for damage from this in North America now (see *River Plants*) and it could be detectable in Britain during the next decade. The signs are:

   Diversity and cover decreased

   Species present have tough and thick parts

   Contrast between affected vegetation in streams with crops close to the water-course banks and normal vegetation where there is a wide strip of 'wild' vegetation by the watercourse

Suitable damage ratings can be devised from this information.

   Dropping or dumping of pesticide and herbicide tins causes serious pollution, especially in dykes and drains (see section 6 on herbicides above).

22. Decreasing edge strips

If crops are planted right up to the watercourse bank, this lessens the possibility of weed control by shading with trees, and increases the chance of fertilisers, pesticides and herbicides reaching the channels (see sections 21 and 21 above).

23. Removing trees

This is usually done to

   Increase access for heavy equipment

   Enable crops to be planted closer to the watercourse (see section 22 above)

   Decrease the work of removing fallen branches, etc.

   Please some planners

For most people, however, trees are aesthetically pleasing along watercourses.

For the use of trees in weed control, see section 7 above.

Rock type map

NOTES FOR USING ROCK TYPE MAP

1. The map on p. 127, of all Britain, is divided into sections. These sections are shown separately on pp. 128–45.

2. As the map is complex, only a few place names could be printed without causing confusion. To make it easier to find rivers, grid lines have been marked on the edges of each page. These should be joined up, using a different colour (e.g. red), and grid references can then be identified.

3. The map shows the major rock types only. On a map of this scale small out-crops have to be omitted. When these are suspected (see Chapters 2, 3 and 4) a large-scale geological map MUST be consulted. A small stream with good quality vegetation, but whose quality does not correspond to that shown under the marked rock type may be on e.g. a band of limestone in a clay area, or of sandstone in a Resistant rock area. Stream character can be completely altered if the stream is sited so that the alien rock forms an important part of its catchment.

4. The rock types are marked according to their effect on vegetation: e.g. Midlands (lowland) Coal Measures affect the rivers mainly through the soft shale beds, and are therefore classed as clay. The Pennant sandstone of South Wales is marked as such: however, the area is so polluted that the difference from the Coal Measures is rarely distinguishable. The alluvium in the Plain of York has been omitted as the region is not flat and does not have alluvial dykes. The northern part of the Trent Plain bears dykes (not, of course, on peat) outside the alluvial area marked.

5. Lowland England north of the Severn–Thames line was affected by the Ice Age, and drift materials, particularly clay, were deposited. The large areas of thick clay (where the clay overrides most of the influence of the solid rock below) are marked. The main sand areas are marked. The two circles show areas where the clay mosaic is too complex to be shown on this map. The zones outside these circles have (lesser) clay mosaics.

   South of the Severn–Thames line drift material is less, and seldom affects streams (e.g. clay-with-flints may occur on hilltops without affecting the chalk streams below).

   In highland Britain north of the Severn–Thames line drift deposits (clay, moraine, etc.) are common in valleys, but they seldom have an effect on the fast-flowing streams coming down from the hills above. On local flatter areas, e.g. Anglesey, Boulder Clay influence can be important. The more recent peaty deposits in moorland and bog areas have far more influence; these are not marked.

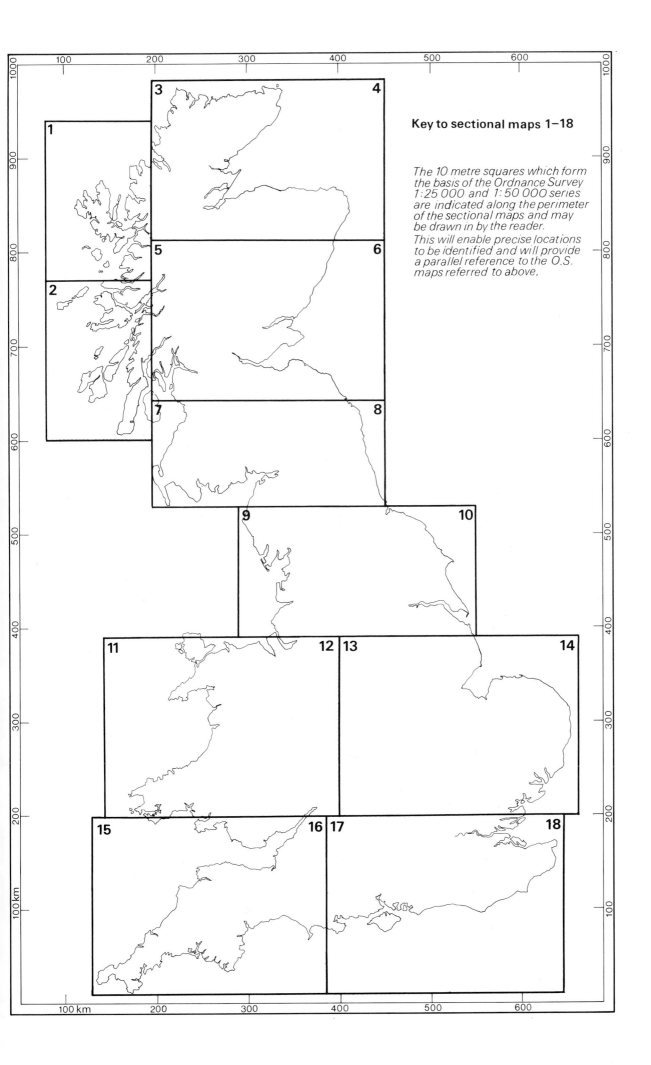

Key to sectional maps 1–18

The 10 metre squares which form the basis of the Ordnance Survey 1:25 000 and 1:50 000 series are indicated along the perimeter of the sectional maps and may be drawn in by the reader.

This will enable precise locations to be identified and will provide a parallel reference to the O.S. maps referred to above.

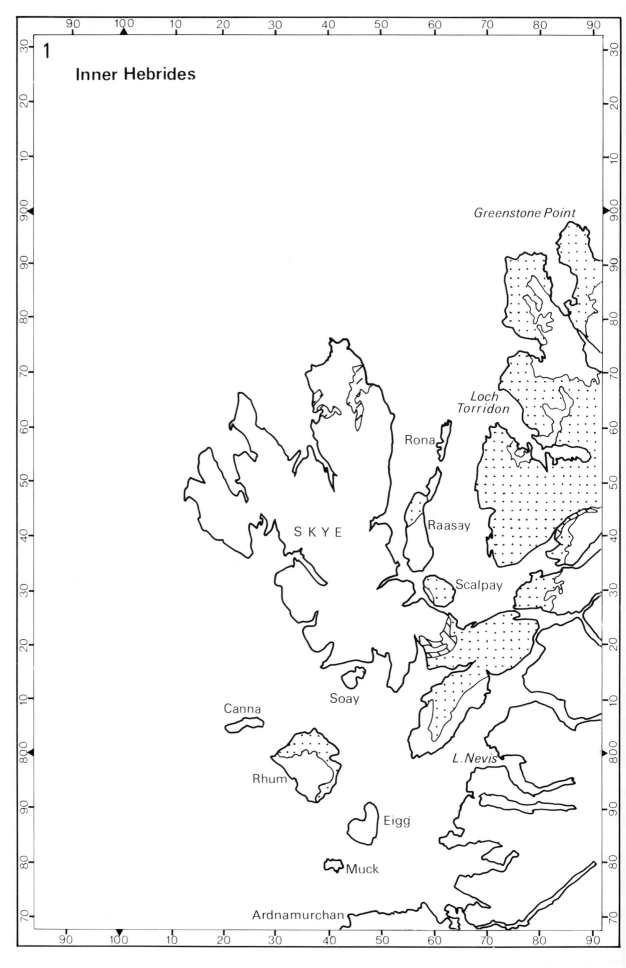

1

Inner Hebrides

Greenstone Point

Loch Torridon

Rona

SKYE

Raasay

Scalpay

Soay

Canna

L. Nevis

Rhum

Eigg

Muck

Ardnamurchan

128

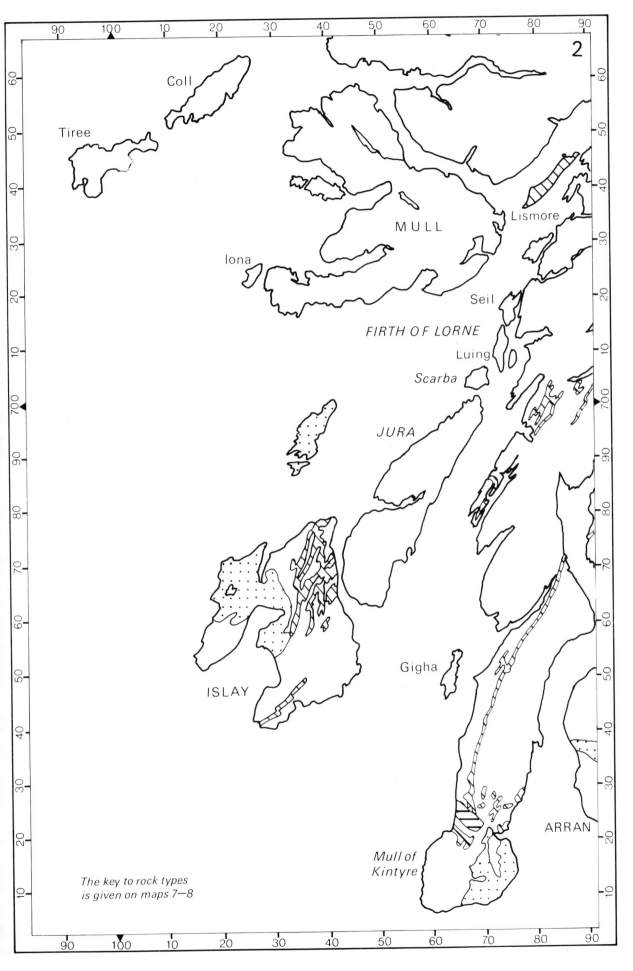

2

Tiree

Coll

M U L L

Lismore

Iona

Seil

FIRTH OF LORNE

Luing

Scarba

J U R A

700

Gigha

ISLAY

ARRAN

*Mull of
Kintyre*

The key to rock types
is given on maps 7–8

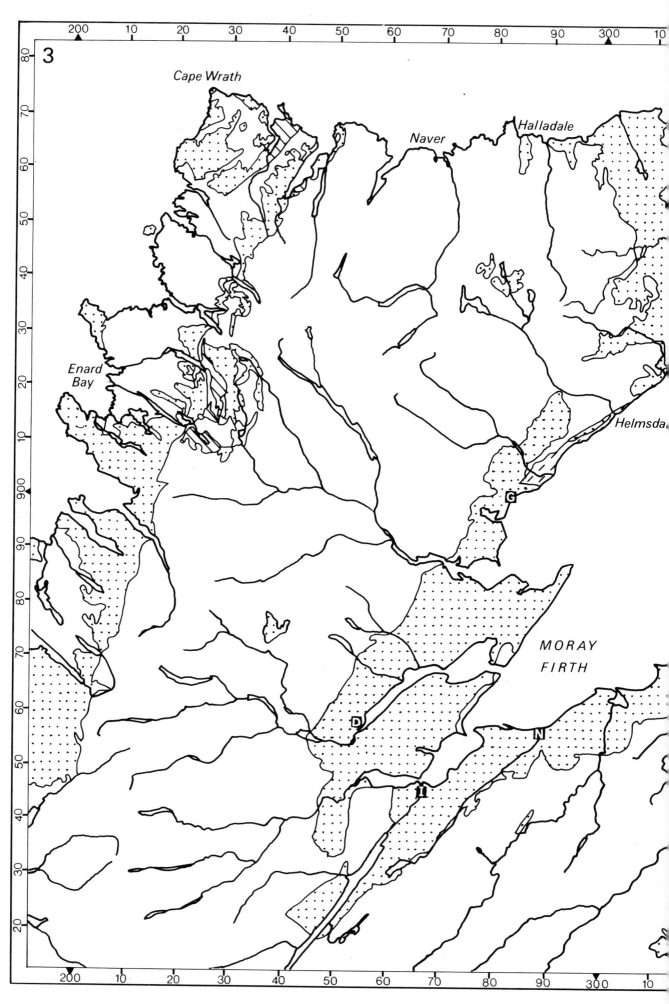

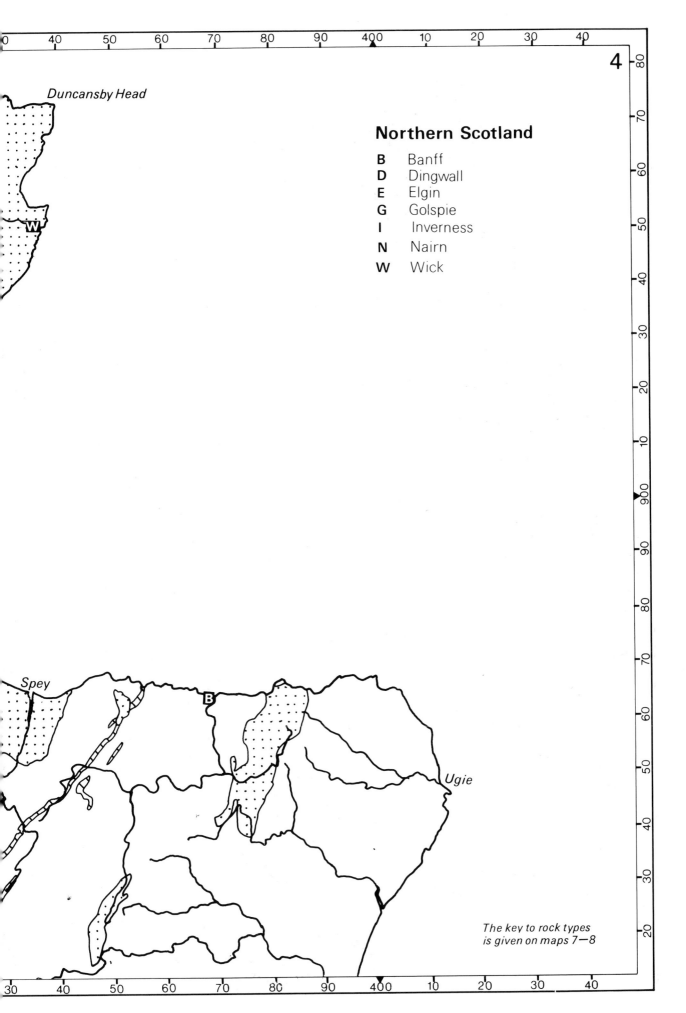

Northern Scotland

B Banff
D Dingwall
E Elgin
G Golspie
I Inverness
N Nairn
W Wick

Duncansby Head

Spey

Ugie

*The key to rock types
is given on maps 7—8*

4

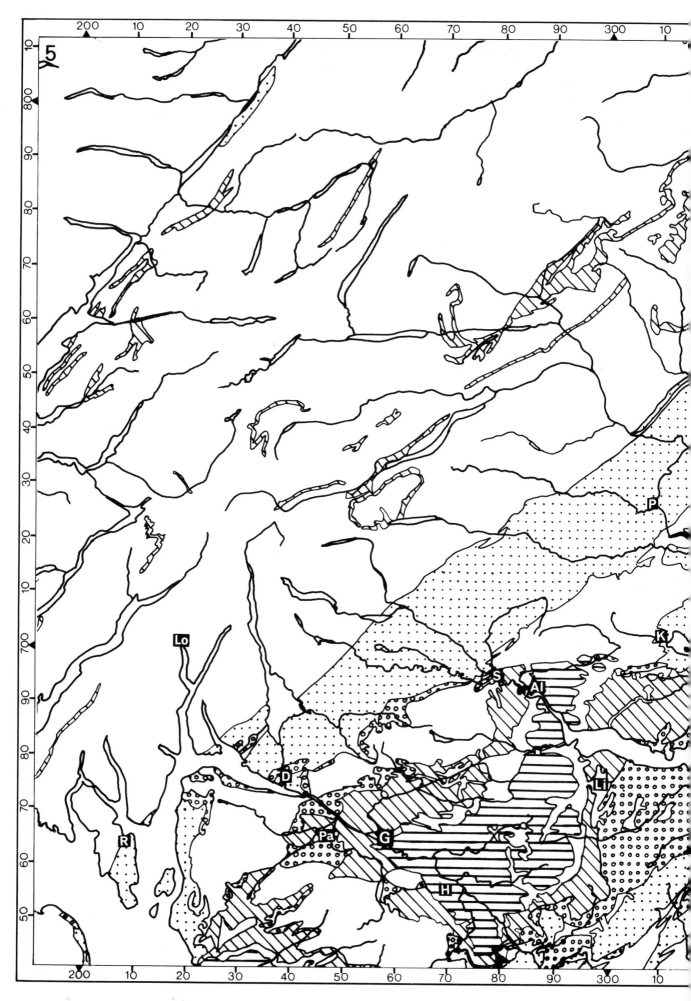

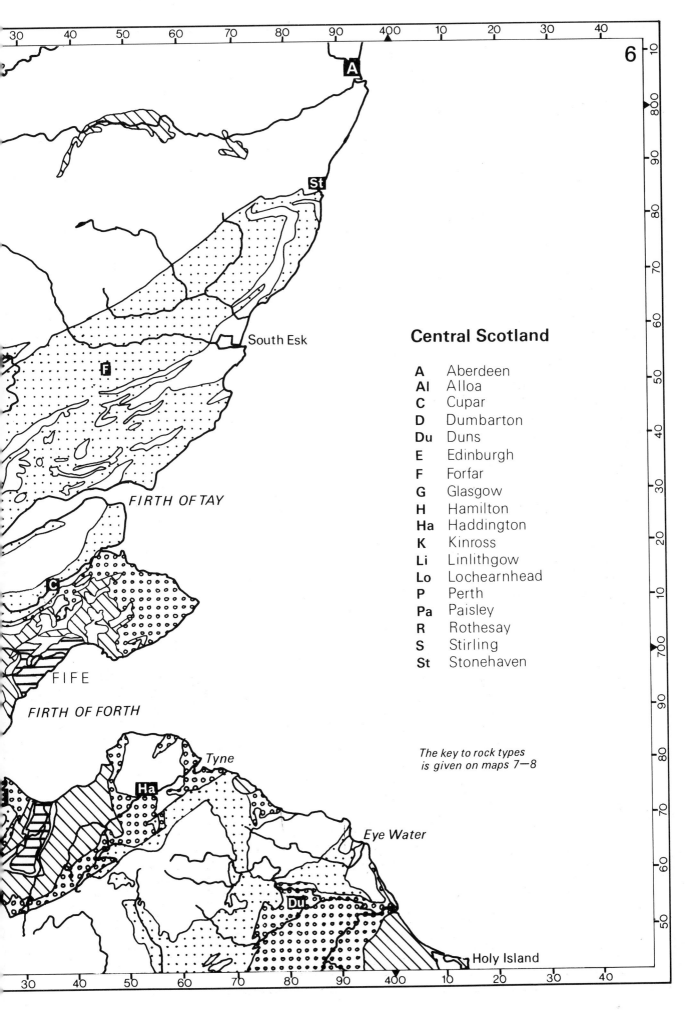

Central Scotland

A	Aberdeen
Al	Alloa
C	Cupar
D	Dumbarton
Du	Duns
E	Edinburgh
F	Forfar
G	Glasgow
H	Hamilton
Ha	Haddington
K	Kinross
Li	Linlithgow
Lo	Lochearnhead
P	Perth
Pa	Paisley
R	Rothesay
S	Stirling
St	Stonehaven

The key to rock types is given on maps 7—8

South Esk

FIRTH OF TAY

FIFE

FIRTH OF FORTH

Tyne

Eye Water

Holy Island

6

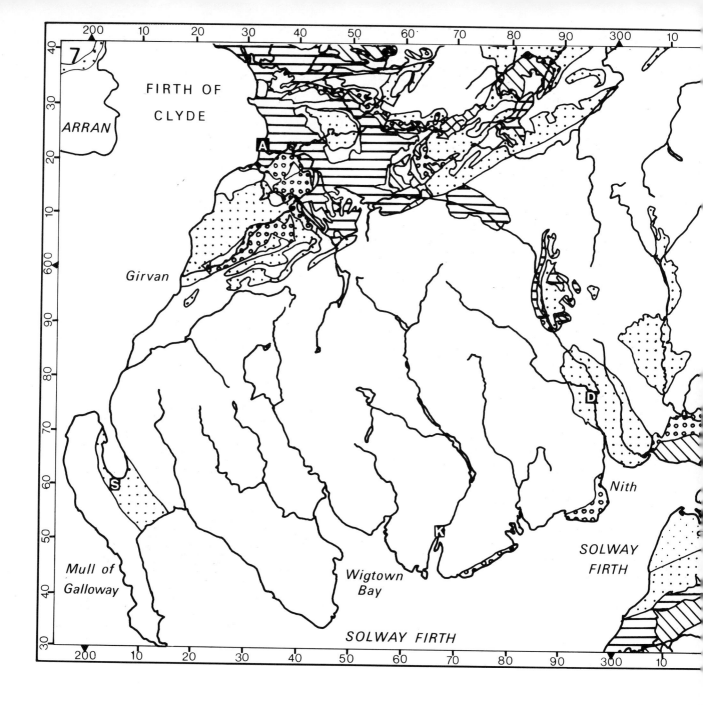

Southern Scotland / Borders

A   Ayr
B   Newtown St Boswells
C   Carlisle
D   Dumfries
Du   Durham
K   Kirkcudbright
N-T Newcastle upon Tyne
P   Peebles
S   Selkirk

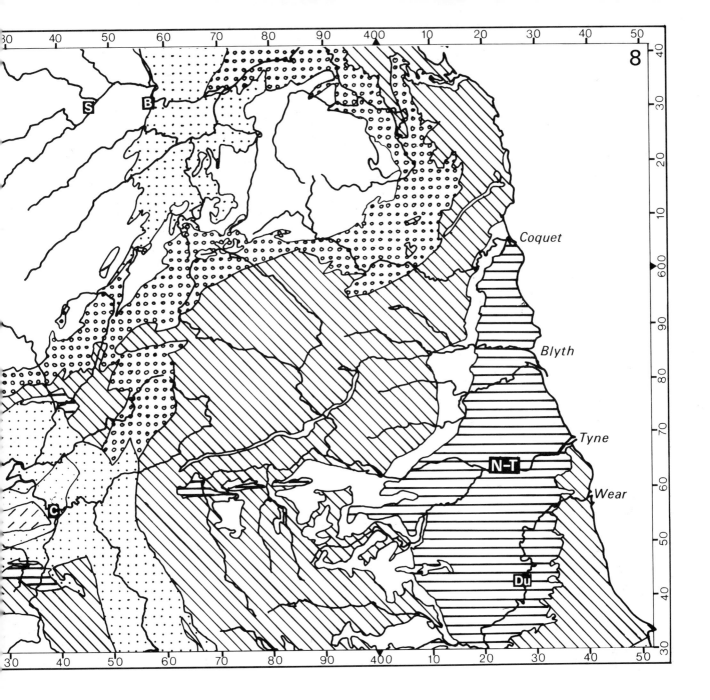

8

KEY

Sandy Drift	Hard Limestone
Boulder Clay	Calciferous Sandstone
Alluvium and Peat	Soft Sandstone
New Forest Sands	Hard Sandstone
Clay	Coal Measures
Chalk	Resistant Rocks
Oolite etc.	

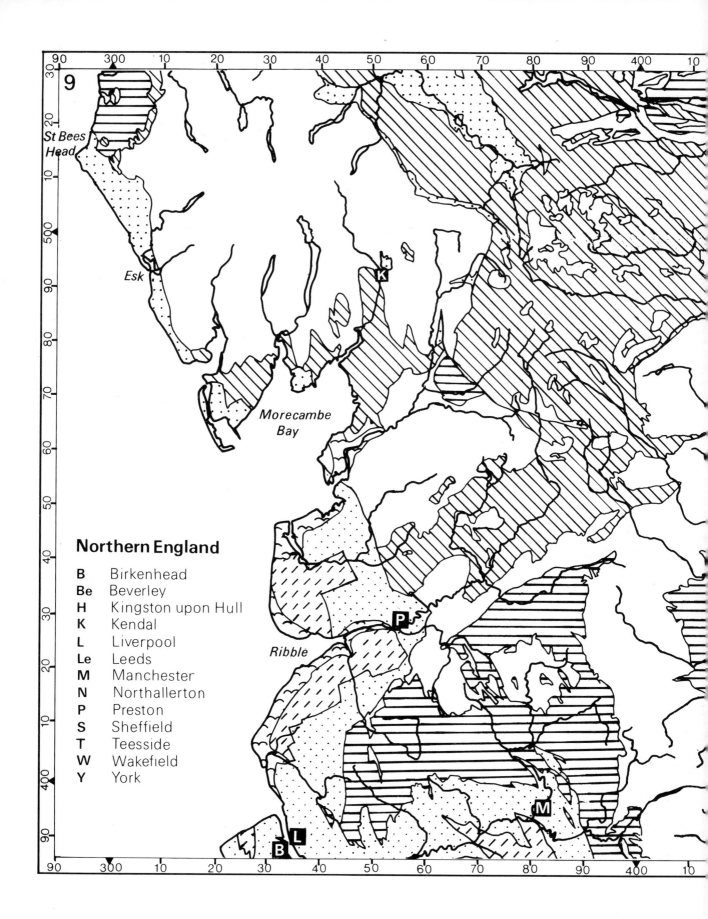

Northern England

B	Birkenhead
Be	Beverley
H	Kingston upon Hull
K	Kendal
L	Liverpool
Le	Leeds
M	Manchester
N	Northallerton
P	Preston
S	Sheffield
T	Teesside
W	Wakefield
Y	York

St Bees Head

Esk

Morecambe Bay

Ribble

9

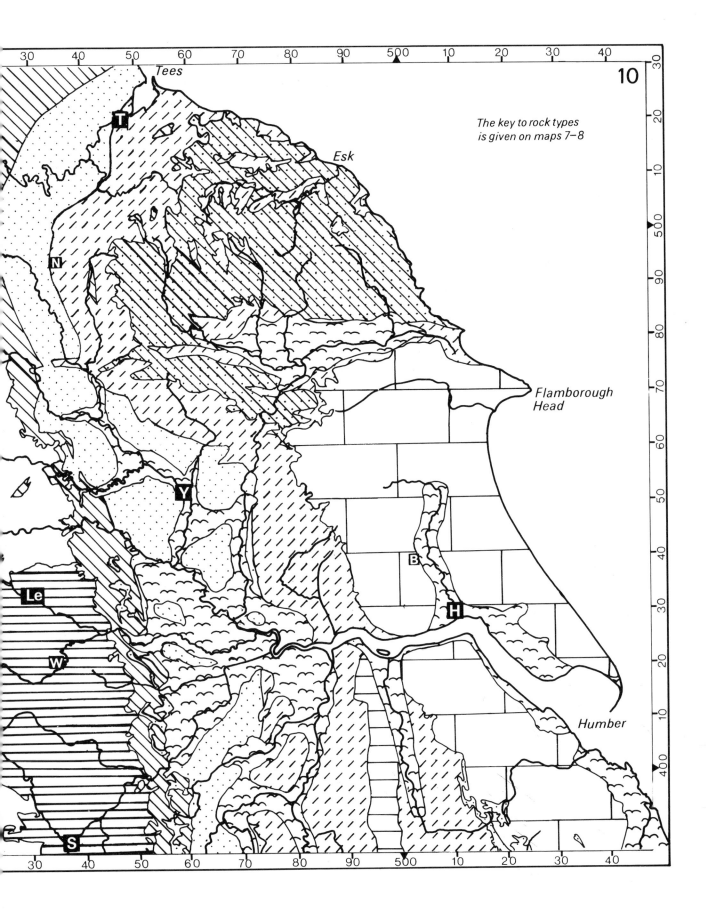

The key to rock types
is given on maps 7–8

Tees

Esk

Flamborough
Head

Humber

T

N

Y

B

H

Le

W

S

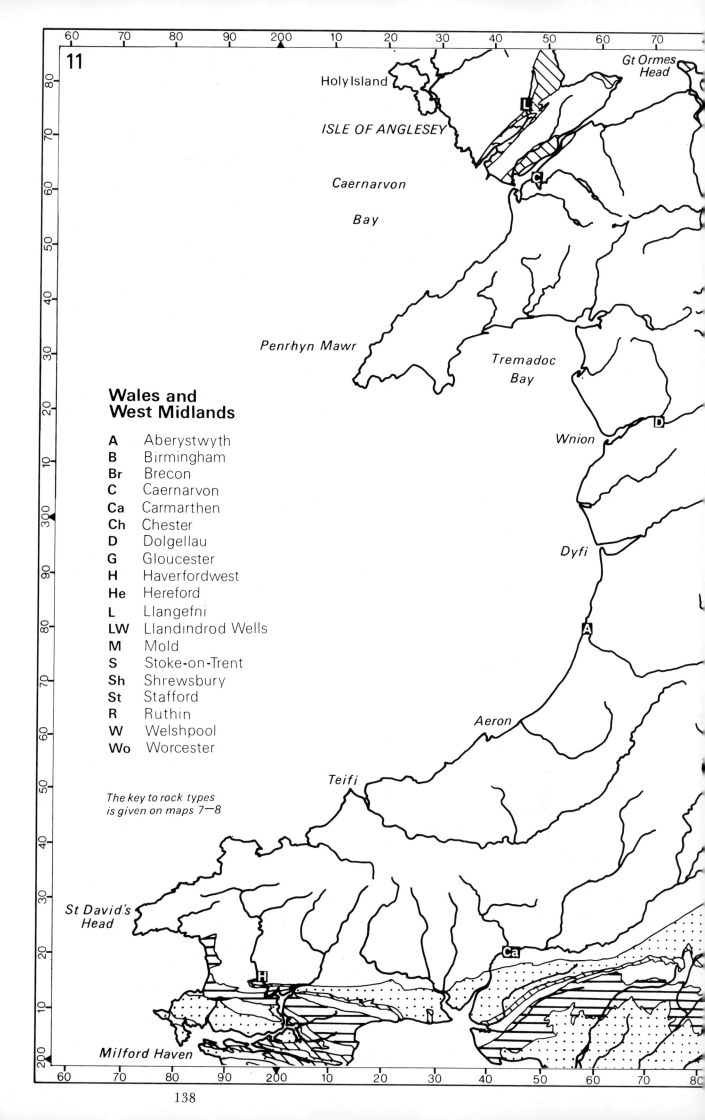

Wales and West Midlands

A   Aberystwyth
B   Birmingham
Br Brecon
C   Caernarvon
Ca Carmarthen
Ch Chester
D   Dolgellau
G   Gloucester
H   Haverfordwest
He Hereford
L   Llangefni
LW Llandindrod Wells
M   Mold
S   Stoke-on-Trent
Sh Shrewsbury
St Stafford
R   Ruthin
W   Welshpool
Wo Worcester

*The key to rock types
is given on maps 7—8*

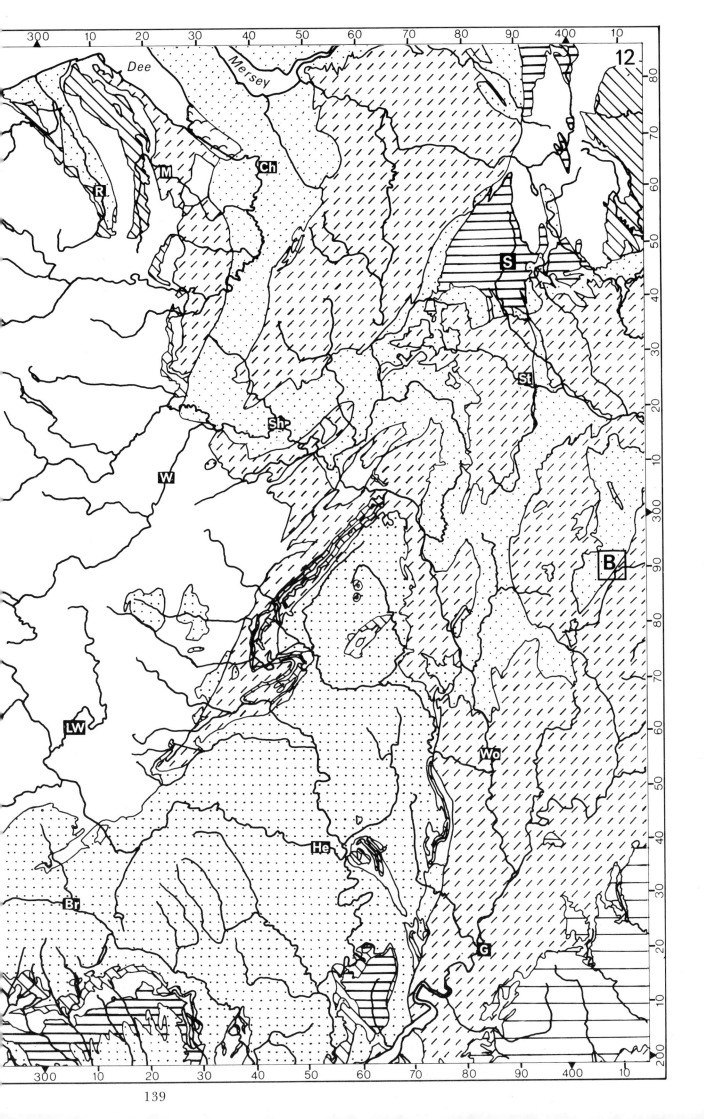

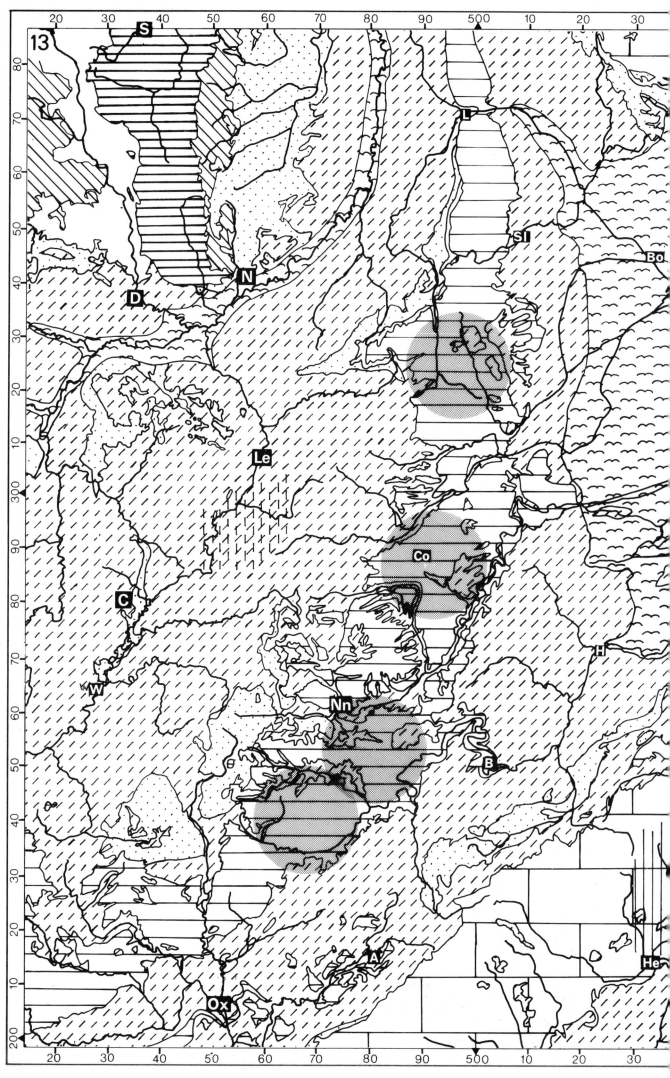

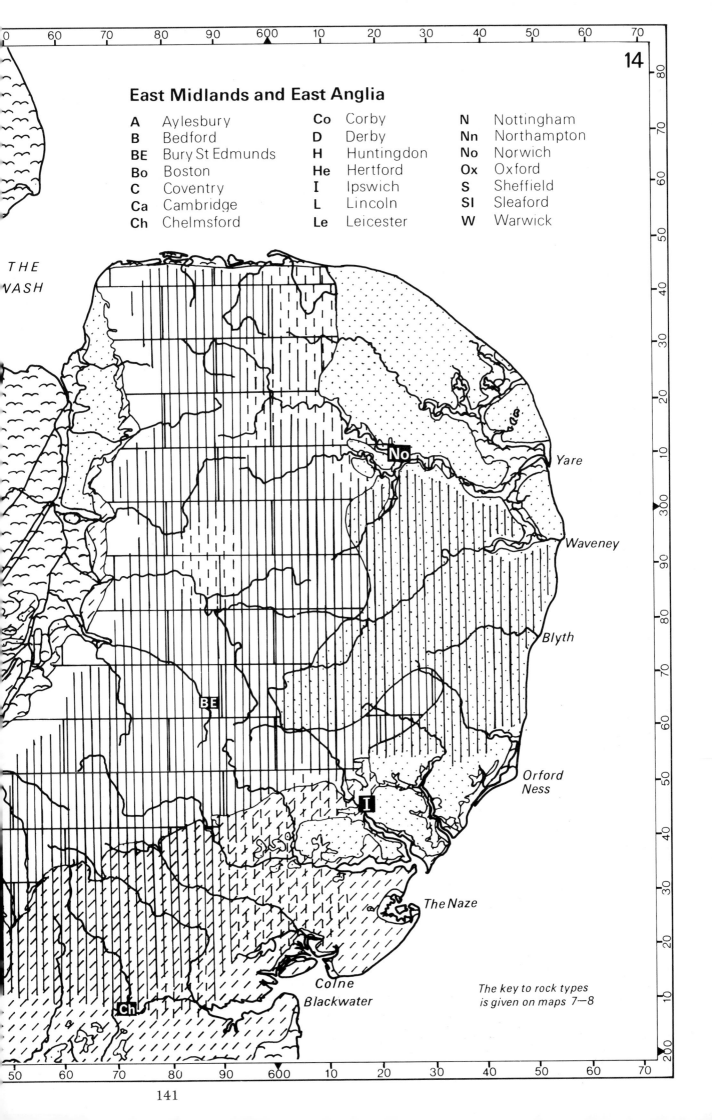

East Midlands and East Anglia

A	Aylesbury	**Co**	Corby	**N**	Nottingham
B	Bedford	**D**	Derby	**Nn**	Northampton
BE	Bury St Edmunds	**H**	Huntingdon	**No**	Norwich
Bo	Boston	**He**	Hertford	**Ox**	Oxford
C	Coventry	**I**	Ipswich	**S**	Sheffield
Ca	Cambridge	**L**	Lincoln	**Sl**	Sleaford
Ch	Chelmsford	**Le**	Leicester	**W**	Warwick

THE WASH

Yare

Waveney

Blyth

Orford Ness

The Naze

Colne
Blackwater

*The key to rock types
is given on maps 7—8*

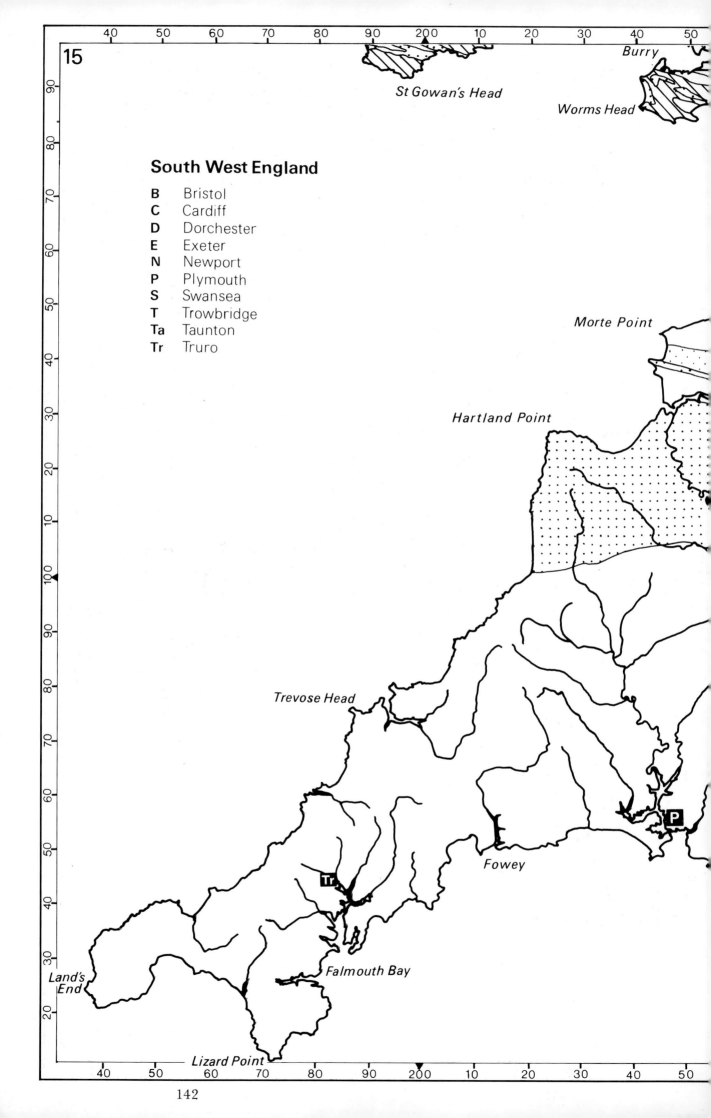

South West England

B   Bristol
C   Cardiff
D   Dorchester
E   Exeter
N   Newport
P   Plymouth
S   Swansea
T   Trowbridge
Ta Taunton
Tr Truro

15

St Gowan's Head

Burry

Worms Head

Morte Point

Hartland Point

Trevose Head

Fowey

Tr

P

Falmouth Bay

Land's End

Lizard Point

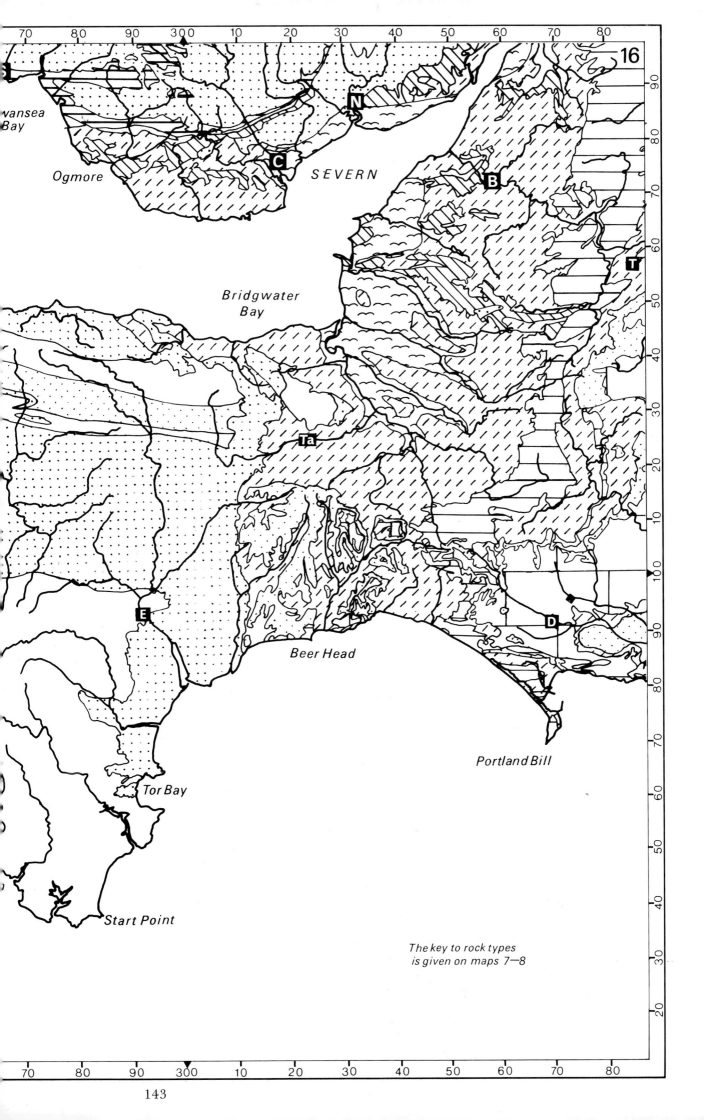

16

vansea
Bay

Ogmore

SEVERN

N

C

B

T

*Bridgwater
Bay*

Ta

E

Beer Head

D

Portland Bill

Tor Bay

Start Point

*The key to rock types
is given on maps 7–8*

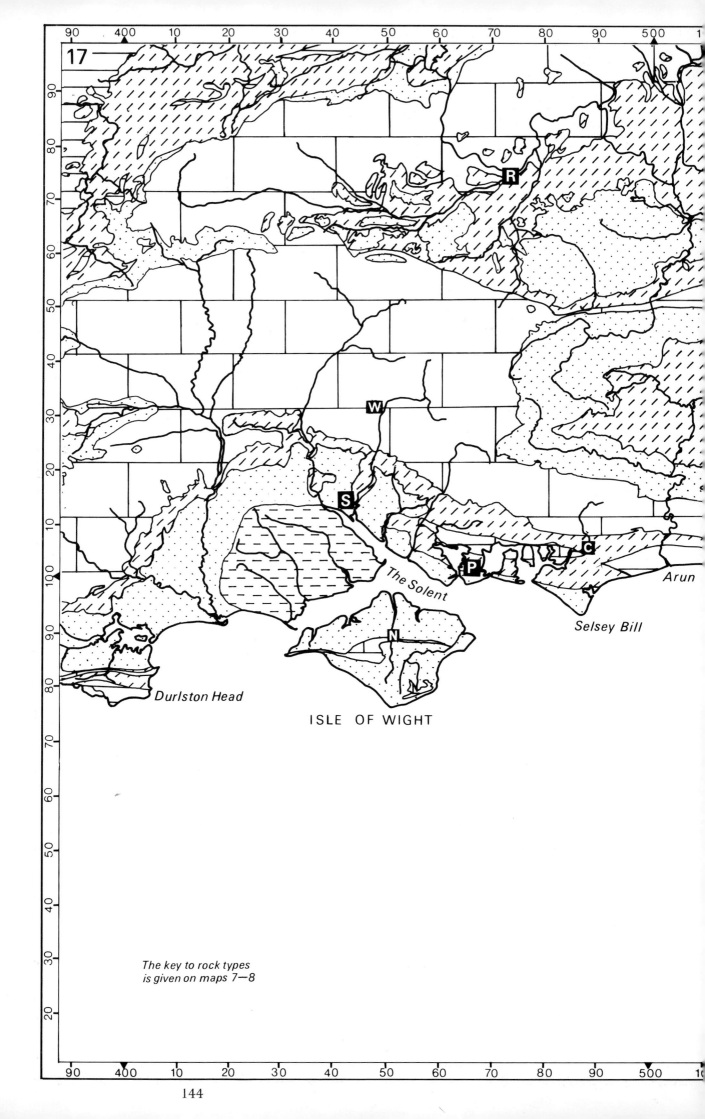

The key to rock types
is given on maps 7—8

Durlston Head

The Solent

ISLE OF WIGHT

Selsey Bill

Arun

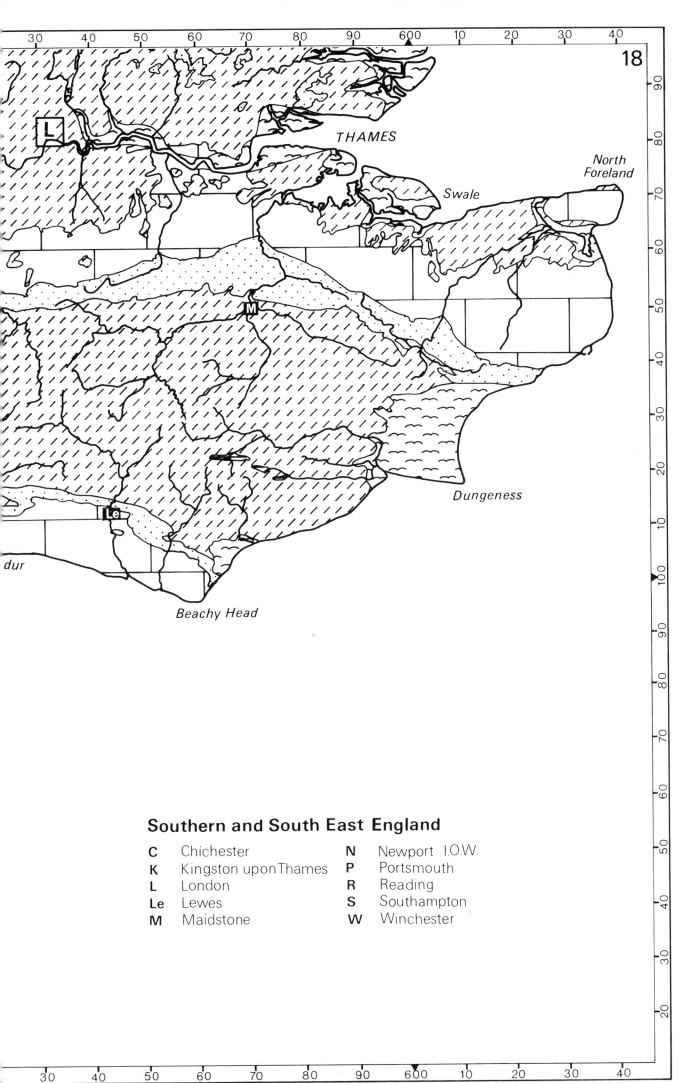

18

THAMES

Swale

North Foreland

M

Le

dur

Dungeness

Beachy Head

Southern and South East England

C   Chichester
K   Kingston upon Thames
L   London
Le Lewes
M   Maidstone

N   Newport I.O.W.
P   Portsmouth
R   Reading
S   Southampton
W   Winchester

Glossary

Acid   Poor in nutrients, of low pH.

Agg. (aggregate)   Used after a plant name to denote an aggregate of species which are difficult to identify separately: e.g. *Rorippa nasturtium-aquaticum* agg. for *Rorippa nasturtium-aquaticum*, *Rorippa microphylla*, and the hybrid between them.

Agriculture   Cultivation (tillage) of the land.

Alluvial plain   Flat tract of country composed of alluvium.

Alluvium   Deposits of silt (sand, etc.) left by water flowing over land which is not permanently submerged; especially those deposits left in river valleys and deltas.

Aquatic   (1) Living or growing in or near water. (2) An aquatic plant or animal.

Aquifer   Rock which yields water.

Arable   Land fit for ploughing and tillage; not grassland, moorland, woodland or built-up land; bearing crops.

Associate species   Species united in the same group or category, subordinate to abundant or dominant species (see below).

Bed (of river)   Bottom or floor of watercourse.

Biomass   Total weight of living things present at any one time.

Blanket bog   Nearly flat tract of country composed of wet bog, wet acid peat.

Blanket weed   Filamentous algae (chiefly *Cladophora*) large enough to trail from the watercourse bed.

Bog   Wet spongy ground, consisting chiefly of decayed moss and other plants; nutrient-poor and acid.

Boulder Clay   A clayey deposit of the Ice Age, containing boulders — but affecting watercourse plants as clay.

Brook   Small stream.

Canal   Artificial watercourse uniting rivers, lakes or seas for the purpose of inland navigation.

Catchment   Natural drainage area or basin, wherein rainfall is caught and channelled to a single exit point (see Watershed).

Chalk   Soft, white, comparatively pure limestone, consisting of calcium carbonate.

Clay   Soft rock, or stiff earth, consisting mainly of aluminium silicate and derived mostly from the decomposition of felspathic rocks.

Clean   Unpolluted.

Coal Measures   The rocks formed by seams of coal and the intervening strata of clay, sandstone, etc.

Community, plant   The plants present in a site and their social ordering.

Cover   Area occupied by vegetation.

Damage   Loss or detriment caused by harm or injury.

Damage rating   Method of assessing damage to watercourse vegetation.

Discharge   Total volume of water per unit time flowing through the channel.

Diversity   Number of species present.

Dominant   Of a plant species, occupying the most space in, and/or controlling the character of, a plant community.

Drain   Drainage channel, the larger channels of the Fenland etc. drainage system, receiving water from dykes, usually by gravity, and passing it to the rivers etc. usually through pumping systems. Usually 6—20 m wide.

Drift, glacial   Material deposited during the glaciations of the Quaternary era (as opposed to the solid rock of more ancient origin).

Dyke   An artificial watercourse for draining marshy land and moving surface water. Usually 2—4 m wide; found beside roads and between fields; draining into larger drains, sewers, rivers or the sea. Named originally in East Anglia and Fenland.

Dystrophic   Of negligible nutrient content; acid and usually composed of, or stained with, bog peat. By extension, used to describe species characteristic of such a habitat.

Ecology   Study of plants and animals in their habitats; mutual relations between plants and animals and their environment.

146

Effluent   Outflow from sewage works, factories, farms, etc.

Emerged (of plant parts)   Above water.

Emergent   A plant mainly or entirely above water.

Ephemeral   Short-lived, transitory.

Erosion   Scour; the removal of material from the channel of a stream.

Eutrophic   Of high nutrient regime, with ample or even excess nutrients for plant growth. By extension, used to describe species characteristic of nutrient-rich habitats.

Eutrophication   Raising of nutrient status.

Fen   Low land, now or formerly covered with shallow water, or intermittently so covered. Any peat developed is alkaline (contrast bog peat) because of the high base-status of the water, derived from the land around. Fen silt is alluvial and nutrient-rich. Particularly the Fenland of East England; by extension, other regions.

Fertile   Nutrient-rich.

Flash flood   Storm flow in which discharge rises very rapidly, due to a combination of heavy rainfall and quick run-off from the catchment.

Floating   Lying flat on and in the plane of the water surface, or occasionally just below it.

Flood   (1) An overflowing of water over land. (2) A storm flow.

Flood gate   Contrivance for stopping or regulating the passage of water.

Flood hazard   That which, by obstructing water movement, may or will cause flooding.

Flow   Water movement; quantity of water moving.

Fringing herbs   Group of semi-emergent, somewhat bushy short dicotyledons, commonly fringing the edge of certain stream types, and occurring more sparsely in a wider range of types.

Gates, flood   See Flood gates.

Grapnel   An instrument for grasping plant parts (see Chapter 1).

Hard rock   Here used to describe those very hard types of rock which do not erode easily, and so can potentially, and often actually, form hilly and mountainous country. They include Carboniferous limestone, Coal Measures, Resistant rocks and Old Red Sandstone.

Hardness ratio   Chemical parameter devised by Dr B. Seddon (1972. 'Aquatic macrophytes as limnological indicators'. *Freshwater Biology*, 2, 107–30). It is calculated as the (calcium plus magnesium) content divided by the (sodium plus potassium) content (here, of silt interstitial water).

Headwater   Stream near the source of a river.

Herbaceous   Plants not forming wood, but dying down every year.

Herbicide   Chemical which kills plants; used for weed control.

Inorganic   Not formed from plant or animal parts (except when these have been completely broken down); mineral.

Irrigation   Supplying land with water.

Landscape   Country scenery.

Limestone   Rock consisting chiefly of carbonate of lime (calcium).

Lode   Navigable watercourse in the Fenland, usually developed from a stream. Commonly 3–10 m wide.

Lowland   Low-lying land. Used here in a specialised sense, defined on p. 14.

Macrophyte   Large plant; the plants discussed in this book, the higher plants (angiosperms) horsetails, water fern, mosses, liverworts and the large algae (e.g. *Chara*, *Cladophora*, *Enteromorpha*).

Mesotrophic   Of moderate nutrient regime. By extension, used to describe species characteristic of these habitats.

Mill (water mill)   Building fitted with machinery for grinding corn, in which the power is provided by a wheel on a stream. Most British mill wheels are disused.

Mill pool   Pool or pond often found in a mill stream somewhat downstream of the mill wheel; sometimes upstream to give a head of water.

Mill stream   Stream turning a mill wheel. Normally is a branch from the main stream, with structures for water regulation.

Mineral   Natural substance of neither animal nor plant origin; inorganic.

Moor   Uncultivated land with some (dry) acid peat or humus and much heather or similar species.

Mountain   Large hill. A mountain stream is here defined in a specialised sense (see p. 14).

Nutrient   Serving as nourishment; normally used of inorganic substances necessary for plant growth, such as calcium phosphate, etc.

Oligotrophic   Low in nutrients. By extension, used to describe species characteristic of this habitat.

Oolite   Pale soft limestone, composed of small rounded granules, with a lower proportion of calcium carbonate than chalk.

Organic   Of, or pertaining to, or composing plants or animals.

Outcrop   The cropping out or exposure of a rock type at the surface.

Particle   Portion of matter.

Peat   Plant material stored and partly decomposed under water. Found in fens (alkaline peat), bogs (acid peat), moors etc. (acid peat), and swamps.

Perennial flow   Of a stream flowing throughout the year.

Pesticide   Chemical which kills pests; usually used for those killing small animals which are dangerous for crop production or human health.

Plain   Flat tract of country.

Pollutant   Substance causing pollution.

Pollution   The alteration of the chemical status of a watercourse by human interference; chemical damage to vegetation.

Pollution index   Method of assessing the effects of pollution on stream vegetation.

Reach   Portion of a river etc. which can be seen in one view. Hence 'lower reaches' for the lower or downstream end of a river and 'upper reaches' for the part nearer the sources.

Reen   See Rhyne.

Resistant rock   Very hard rock resistant to both erosion and solution; includes andesite, basalt, gneiss, granite, Millstone Grit, schist, shale and slate.

Rhyne (reen)   Name given to a system of dykes and drains in South Wales and south-west England.

Rill   A small brook, rivulet.

River   A large stream of water flowing in a channel towards the sea, a lake, or another stream.

Rock (bedrock)   Material composing the hard surface of the earth, e.g. clay, limestone, Resistant rock, sandstone.

Rock (of particle size)   Bedrock exposed in the channel, or large particles of the size of boulders.

Run-off   Water flowing off the land into watercourses etc.

Sandstone   Rock composed of consolidated sand.

Scour   The action of a current or flow of water in clearing away silt and other sediments.

Sediment   Particles which fall by gravity in water; mud, silt, sand, gravel, stones and boulders.

Sewer   In southern England, an artificial watercourse for draining marshy land and moving surface water into a river or the sea. Often 2–6 m wide.

Silting   Depositing silt.

Soft rock   Here used to describe the softer types of rock which erode easily and form lowland landscape; e.g. chalk, oolite, clay (mostly), Tertiary sandstones, etc.

Solid rock   All rock types except Glacial Drift and recent alluvial deposits.

Solute   A dissolved substance.

Spate   Large discharge or storm flow caused by heavy rains, etc., in hill streams where the water force is great.

Species   Group of plants (or animals) having certain common and permanent characteristics distinguishing it from other groups.

Spring   Flow of water rising or issuing naturally out of the ground.

Storm flow   The large water discharge that follows heavy rain.

Stream   Course of water flowing continuously along a bed on the earth, forming a river or brook.

Stress   Pressure of some adverse force or influence.

Structure   Used for all constructions altering flow in streams, e.g. lock, weir, sluice.

Submerged (of plant parts)   Within the water.

Submergent   A plant within the water.

Subsoil   Stratum of soil lying immediately under the surface soil.

Substrate   Material near the surface of the bed of the watercourse; the rooting medium; the soil.

Tall monocotyledons   Group of tall emergent aquatics with long narrow leaves, forming dense stands which shade out all shorter plants.

Topography   Features of a region or locality.

Toxic   Poisonous.

Trophic   Of or pertaining to nutrition.

Turbid   Thick or opaque with suspended matter; not clear; cloudy.

Turbulence   Agitation, disturbance, commotion of the water.

Undamaged (vegetation)   Here plant communities with the highest category of diversity and cover now found in unpolluted watercourses.

Unstable   Apt to change or alter.

Upland   Hilly country. An upland stream is here defined in a specialised sense (see p. 14).

Vegetation   Plants in general; the plant life at a site.

Very mountainous   Steep and hilly country. A very mountainous stream is here defined in a specialised sense (see p. 14).

Glossary

Watercourse   A stream of water, a river or brook; an artificial channel for the movement of
  water. The general term for water channels, including all the other types defined here.
Watershed   Tract of ground between two drainage basins.
Water-supported   Of plants supported by the water: floating and submerged species.
Water transfer   Movement of water, usually for domestic supply, from one river to another.
Weir   Barrier or dam to restrain water.

Index to sections

(For reference to sections on identification and assessment only.)

List of abbreviations
of plant names on stream dial

BROWN

Sphag	*Sphagnum* spp.
Dros rot	*Drosera rotundifolia*
Dros angl	*Drosera anglica*
Narth oss	*Narthecium ossifragum*
Erioph ang	*Eriophorum angustifolium*
Litt unif	*Littorella uniflora*
Meny tri	*Menyanthes trifoliata*
Pot polyg	*Potamogeton polygonifolius*
Ran flam	*Ranunculus flammula*
Sparg ang	*Sparganium angustifolium*
Pot gram	*Potamogeton gramineus*

ORANGE

Calth pal	*Caltha paulstris*
Ran hed	*Ranunculus hederaceus*
Ran omio	*Ranunculus omiophyllum*
Oen croc	*Oenanthe crocata*
Myr alt	*Myriophyllum alterniflorum*
Call ham	*Callitriche hamulata*
Junc art	*Juncus articulatus*
Eleoch acic	*Eleocharis acicularis*
Eleog fluit	*Eleogiton fluitans*
Junc bulb	*Juncus bulbosus*
Nymph alb	*Nymphaea alba*
Nuph lut	*Nuphar lutea*
Eleoch pal	*Eleocharis palustris*

YELLOW

Glyc fluit	*Glyceria fluitans*
Pot sparg	*Potamogeton × sparganifolius*
Car acuta	*Carex acuta*
Pot alp	*Potamogeton alpinus*
Phal arund	*Phalaris arundinacea*
Mosses	Mosses
Ran aq	*Ranunculus aquatilis*
Ran pen	*Ranunculus penicillatus*
Car acutif	*Carex acutiformis* agg.
Bl wd	Blanket weed (trailing algae, mainly *Cladophora*)

GREEN

Pet hyb	*Petasites hybridus*
Polyg amph	*Polygonum amphibium*
Pot nat	*Potamogeton natans*
Ran fluit	*Ranunculus fluitans*

BLUE

Mim gutt	*Mimulus guttatus*
L tri	*Lemna trisulca*
Ver becc	*Veronica beccabunga*
Ber erect	*Berula erecta*
Ran aq	*Ranunculus aquatilis*
Ran pelt	*Ranunculus peltatus*
Call spp. (not ham)	*Callitriche* spp. (not *hamulata*)
Mosses	Mosses
Ment aq	*Mentha aquatica*
Ror nast-aq	*Rorippa nasturtium-aquaticum* agg.
Catab aq	*Catabrosa aquatica*
Ran calc	*Ranunculus calcareus*
Ran pen	*Ranunculus penicillatus*
Ran trich	*Ranunculus trichophyllus*
Sol dulc	*Solanum dulcamara*
Ap nod	*Apium nodiflorum*

Myos scorp	*Myosotis scorpioides*
Ver anag-aq	*Veronica anagallis-aquatica* agg.
Hipp vulg	*Hippuris vulgaris*
Sparg erect	*Sparganium erectum*

PURPLE

Elod can	*Elodea canadensis*
Car acutif	*Carex acutiformis* agg.
T lat	*Typha latifolia*
Phrag com	*Phragmites communis*
Pot perf	*Potamogeton perfoliatus*
Ran trich	*Ranunculus trichophyllus*
Ran pen	*Ranunculus penicillatus*
Bl wd	Blanket weed
Groenl dens	*Groenlandia densa*
Zann pal	*Zannichellia palustris*
Oen fluv	*Oenanthe fluviatilis*
Glyc max	*Glyceria maxima*
Alisma pl-aq	*Alisma plantago-aquatica*
Pot luc	*Potamogeton lucens*
Myr spic	*Myriophyllum spicatum*
Pot crisp	*Potamogeton crispus*
Phal arund	*Phalaris arundinacea*

RED

Epil hirs	*Epilobium hirsutum*
Entero	*Enteromorpha* sp.
Cerat dem	*Ceratophyllum demersum*
Rum hydr	*Rumex hydrolapathum*
Sparg emer	*Sparganium emersum*
Sag sag	*Sagittaria sagittifolia*
But umb	*Butomus umbellatus*
Pot pect	*Potamogeton pectinatus*
Ror amph	*Rorippa amphibia*
Sch lac	*Schoenoplectus lacustris*
Nuph lut	*Nuphar lutea*

WHITE

L mi	*Lemna minor* agg.
Agr stol	*Agrostis stolonifera*